Oriol Costa Echaniz

The role of sustainable consumption in the smart sustainable cities

Oriol Costa Echaniz

The role of sustainable consumption in the smart sustainable cities

LAP LAMBERT Academic Publishing

Impressum / Imprint

Bibliografische Information der Deutschen Nationalbibliothek: Die Deutsche Nationalbibliothek verzeichnet diese Publikation in der Deutschen Nationalbibliografie; detaillierte bibliografische Daten sind im Internet über http://dnb.d-nb.de abrufbar.
Alle in diesem Buch genannten Marken und Produktnamen unterliegen warenzeichen-, marken- oder patentrechtlichem Schutz bzw. sind Warenzeichen oder eingetragene Warenzeichen der jeweiligen Inhaber. Die Wiedergabe von Marken, Produktnamen, Gebrauchsnamen, Handelsnamen, Warenbezeichnungen u.s.w. in diesem Werk berechtigt auch ohne besondere Kennzeichnung nicht zu der Annahme, dass solche Namen im Sinne der Warenzeichen- und Markenschutzgesetzgebung als frei zu betrachten wären und daher von jedermann benutzt werden dürften.

Bibliographic information published by the Deutsche Nationalbibliothek: The Deutsche Nationalbibliothek lists this publication in the Deutsche Nationalbibliografie; detailed bibliographic data are available in the Internet at http://dnb.d-nb.de.
Any brand names and product names mentioned in this book are subject to trademark, brand or patent protection and are trademarks or registered trademarks of their respective holders. The use of brand names, product names, common names, trade names, product descriptions etc. even without a particular marking in this work is in no way to be construed to mean that such names may be regarded as unrestricted in respect of trademark and brand protection legislation and could thus be used by anyone.

Coverbild / Cover image: www.ingimage.com

Verlag / Publisher:
LAP LAMBERT Academic Publishing
ist ein Imprint der / is a trademark of
OmniScriptum GmbH & Co. KG
Heinrich-Böcking-Str. 6-8, 66121 Saarbrücken, Deutschland / Germany
Email: info@lap-publishing.com

Herstellung: siehe letzte Seite /
Printed at: see last page
ISBN: 978-3-659-66400-7

Copyright © 2014 OmniScriptum GmbH & Co. KG
Alle Rechte vorbehalten. / All rights reserved. Saarbrücken 2014

What is the role of sustainable consumption in the smart sustainable cities' projects across Europe?

Oriol Costa Echaniz

Stockholm 2014

Executive Summary

Current projections indicate that by 2050, two in every three people will live in urban areas, and that cities will accommodate 3 billion people during this period. Cities are consuming three-quarters of the world's energy and causing three-quarters of global pollution. To reduce these impacts, new technologies have been considered in the development of smart sustainable cities, but technology has not always favoured the idea of sustainable consumption. To address this issue, we have aimed to focus on identifying the role of sustainable consumption within implementations of smart cities' projects across Europe.

We have selected a set of smart city projects in 76 cities in Europe from CONCERTO initiatives, Mapping Smart Cities in Europe, Energy Study for the Stockholm Region and Joint European Support for Sustainable Investment in City Areas and classified them according to: smart governance, smart mobility, smart living, smart environment, smart citizens and smart economy. Furthermore, we established a number of categories for the classification of the evaluated projects based on their relevance to sustainable consumption, and considered several solutions for the integration of sustainable consumption in smart sustainable cities.

The results show that in 18.9% of the projects, sustainable consumption is not relevant at all. The second classification shows the percentage of the remaining categories where sustainable consumption is relevant; 8.3% consider sustainable consumption as relevant even though it was not implemented in the project. These cities aim to achieve a higher level of sustainable consumption, which is expected to be included in future projects. If they keep themselves in this category, their behavioural consumption patterns will not change and the impact of citizens on the cities will remain the same. The majority of the projects, 54.2%, implemented technology to reduce consumption but if the projects do not coincide with the behaviour of citizens, a big rebound effect will occur. 37.5% of the projects consider relevant sustainable consumption to its full potential and this can change citizen's behaviour.

In conclusion, sustainable consumption is relevant in most of the projects analysed, with new technologies available to help energy savings and reduction of our consumption. However, if there is a lack of smart consumption from the citizens, the technologies available might not be sufficient and consumption could increase. One quarter of the analysed smart cities projects still do not consider the consumption behaviour of the citizens. This can be changed through campaigns and explanations targeting the population on how to manage and reduce energy and resource consumption. To reduce the negative impact of the cities' growth, projects considering smart sustainable cities need to integrate sustainable consumption policies that account for citizens' behaviour.

Keywords: smart sustainable cities, sustainable consumption, rebound effect, cities impact.

Acknowledgement

This thesis has been conducted during the spring semester 2014 at KTH Royal Institute of Technology, Sweden. I would like to thank everyone who has been involved in making the project possible.

In particular, I would like to thank my supervisor, Olga Kordas, and her colleagues for their support and guidance throughout all phases of the thesis. I would like to show my appreciation for their continued effort and patience while helping me define the study. I would also like to thank all the people that work in the Department of Industrial Ecology at KTH.

I am grateful for the opportunity provided by UPC for the study period abroad and particularly my supervisor in Barcelona, Jordi Segalàs, for helping me during the several steps of the thesis and for continually giving me the incentive to work hard.

I would like to thank my friends from around the world, the master (UPC), Barcelona, Sant Cugat, Sweden and Gósol, for their continued support during this long process and their encouragement me to overcome barriers. A special thank you goes to Laura Cutando for the trust me in everything I have done.

Many thanks to fellow housemates for making the long winter days more pleasant and making me feel at home, especially Sara and Paulo for giving me strength and support wherever they were.

Finally, I would like to thank all my family and, in particular, my lovely parents, Elena and Josep, and my brothers, Xavier and Ignasi without them, this experience would not have been possible.

Table of contents

Executive Summary .... I.
Acknowledgement .... II.
Table of contents .... III.
List of figures .... IV.
List of tables .... V.
List of acronyms & abbreviations .... VI.
1. Introduction 1
   1.1 Urbanization of global population 1
   1.2 Evolution of city concepts 2
   1.3 Aims and objectives 4
      1.3.1 Aims 4
      1.3.2 Objectives 4
2. Methodology 6
   2.1 Literature review 6
   2.2 Analysis of projects in European Union 6
   2.3 Evaluation of projects regarding sustainable consumption 6
3. Background 8
   3.1 Sustainable Consumption 8
   3.2 Sectors of Smart Cities 10
      3.2.1 Smart Economy 10
      3.2.2 Smart Mobility 11
      3.2.3 Smart Environment 11
      3.2.4 Smart Society 12
      3.2.5 Smart Living 12
      3.2.6 Smart Governance 13
4. Results 14
   4.1 Definitions of Smart Cities 14
   4.2 Definitions of Smart Sustainable Cities 16
   4.3 Analysis of the smart sectors 20
   4.4 Relevance of sustainable consumption 27
4. Discussion and conclusions 35
5. References 38
Appendix I: Classification of the evaluated projects within smart city sectors and sustainable consumption 41

List of figures

Figure 1: Urban and rural population (United Nations, 2010) 1
Figure 2: The impact of climate polices on greenhouse gas emissions for the EU-27 (Roelfsema et al., 2014) 3
Figure 3: Classification of the different sectors in the projects evaluated 21
Figure 4: Comparison between Concerto's database and Mapping Smart Cities in Europe 22
Figure 5: Percentage of Smart Governance projects in Europe 22
Figure 6: Percentage of Smart Mobility projects in Europe 23
Figure 7: Percentage of Smart Living projects in Europe 24
Figure 8: Percentage of Smart Environment projects in Europe 25
Figure 9: Percentage of Smart Society projects in Europe 26
Figure 10: Percentage of Smart Economy projects in Europe 27
Figure 11: Percentage of the projects evaluated in sustainable consumption categories .. 29
Figure 12: Projects evaluated in Europe with the Category A (SCo not relevant) 30
Figure 13: Projects involved in the B, C and D categories of sustainable consumption in Europe 31

List of tables

Table 1: Fragment of the projects' classification to illustrate how it was organized........... 21

Table 2: Percentage of sustainable consumption categories per sector of smart cities' projects evaluated 28

Table 3: Projects' classification by categories of sustainable consumption relevant 33

List of acronyms & abbreviations

ICT	Information and Communictaions Technology
SC	Smart City
SSC	Smart Sustainable City
SCo	Sustainable Consumption
SD	Sustainable Development
SG	Smart Governance
SM	Smart Mobility
SL	Smart Living
SE	Smart Environment
SS	Smart Society
SEc	Smart Economy
GDP	Gross Domestic Product
EU	European Union
R&D	Research and Development
GPS	Global Positioning System

1. Introduction

1.1 Urbanization of global population

*"For the first time in history, the majority of the world population lives in **cities** and the cities' populations are increasing rapidly"* (Buhaug et al., 2013). One of the main causes of this significant increase in the population of the cities is the migration from the countryside (Buhaug et al., 2013).

Residents of countryside villages generally move to main towns I search of better opportunities and facilities that cities can offer them. Cities have a wide range of opportunities to offer, which make people feel more comfortable and active due to the amount of job on offers, a connection to other parts of the world, research groups and access to foreign markets. Furthermore, cities are places of wisdom and creation *(Egger 2006)*. People want to live close to the social equipment such as schools or hospitals and to be involved in the city's network.

In recent years, several studies have been conducted considering the distribution of the population over the next 30 years in rural and urban areas. *"By 2050, current projections indicate that two in every three people will live in **urban areas**"* (Buhaug et al., 2013). This phenomenon is exemplified in Figure 1. The graph shows how migration to the cities from rural areas increased very fast in a span of a few years, and the projected growing tendency in years to come.

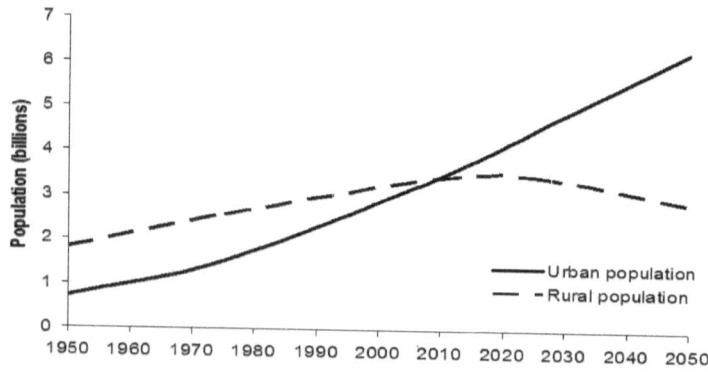

Figure 1: Urban and rural population (United Nations, 2010)

A visible in Figure 1, society should rethink the approaches to the opportunities and challenges of building cities. Both the government and society have to work together and make an effort to think more about long term projects and less about short term profits. We should think more about cooperative strategies rather than competitive ones. If we think short term only, the projects will ultimately be less successful in the long run.

1.2 Evolution of city concepts

*"Years ago, the concept of city said little about the need to be smart and was much less linked to the concept of **sustainability**" (Kramers et al. 2014).*

Cities are characterized by a chain of networked elements. The services, citizens, businesses, transport and communication all interact cooperatively. However, with increased population indexes, needless and easily solvable problems have appeared which are challenging cities. Citizens should expect a supply of public services such as a clean water supply, health care, housing, education, electricity, etc. *(Neirotti et al., 2014).*

The concept of cities has an impact across the globe. *"Cities are consuming three-quarters of the world's energy and causing three-quarters of global pollution" (Rogers, 1998).* Due to the large **impact** generated by the cities, either by the environmental pollution, energy consumption or huge waste generation, it is time to look for solutions to minimize impacts in all areas. *"The impact of cities is determined by a number of factors such as the location, size and population. But much of the impacts are determined by the pattern of life that the citizens have" (Lindberg et al., 2013).* This means the model transport, the **consumption** they generate, the technology they use, etc.

In recent decades, people have changed their habits and have improved their model of life around **new technologies**. *"The usage of the technologies generates a high consumption" (Lindberg et al., 2013).* Some examples would be the lighting, heating, private vehicles, and more recently the use of mobiles phones and computers *(Lindberg et al., 2013 and Herring, 2004).* When a project is completed, it is important to ask whether or not the new technologies that improve the **efficiency** of the products and systems consuming energy also require less consumption from the citizens which can further lessen the environmental impacts.

Nowadays, there have been many conferences on how to make a city a pleasant and sustainable environment. *"After several discussions about that topic, the **smart city** concept was born" (Neirotti et al., 2014).*

When we talk about the evolution of cities in Europe it is necessary to consider the Europe 2020 strategy. *"This is the EU strategy to boost economic growth and job creation in an intelligent and sustainable way" (Manville et al., 2014).* It has established goals to be achieved by 2020 in Europe found in five different areas. *"These areas are climate change and energy, employment, education, R&D and innovation and poverty and social exclusion" (Manville et al., 2014).* In summary, the main points are;

- Climate change and energy: 20% reduction of greenhouse gas emissions in comparison to 1990.
- Employment: 75% of people who are between 20 and 64 years of age must be employed.
- Education: reduce school dropout rates to less than 10%. At least 40% of young people aged 30 to 34 should complete the third level education.
- R&D and innovation: a requirement to invest 3% of EU's GDP in R&D or innovation.
- Poverty and social exclusion: the number of people that are living in poverty must be reduced by at least 20 million.

The impact of climate policies on greenhouse gas emissions for the EU-27 between the years 1990 and 2020 is displayed in Figure 2. According to the European Environment Agency, the European Union emission level was approximately 4.6 gigatonnes of carbon dioxide equivalent in 2011.

It is important to take into account the fact that many current policy initiatives will be implemented by 2020. Furthermore, *"planning is needed on post-2020 policies to achieve the objectives proposed in 2050" (Ruester et al., 2014).*

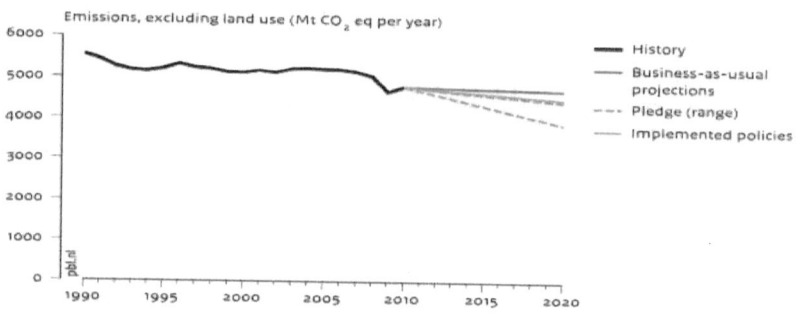

Figure 2: The impact of climate polices on greenhouse gas emissions for the EU-27 (Roelfsema et al., 2014)

Through the Smart Cities implementation it is possible to look at different initiatives to achieve the objectives of the Europe 2020 strategy. Within the area of Smart Cities, there are different sectors where the projects are classified. These sectors are smart governance, smart mobility, smart living, smart environment, smart citizens and smart economy.

Under the term of 'Smart Cities', we find the concept of **ICT** (Information and Communications Technology). These are used to build the services and infrastructure of the city *(Nam and Prado, 2011)*. *"ICTs are very present in the projects undertaken for the evolution of cities and it has the capacity to change society completely" (Kramers et al., 2014).* ICTs help cities improve innovations to achieve better performance and greater efficiency in sectors such as transport, environment, energy, education, health care and safety *(Nam and Prado, 2011)*. *"ICTs are a great solution for the development of sustainability in cities and to minimize the impact of them" (Kramers et al., 2014).* Besides innovating using the best technology, telecommunication companies have to bear in mind what people need, want and expect from new technologies *(Kramers et al., 2014)*. Furthermore, it is important to implement the ICT's in all sectors where sustainability is the focus throughout.

Nowadays, cities want to achieve another goal, and that is to progress from a **smart city** to a **smart sustainable city**. This objective is along the same path as smart cities, but more ambitious. ICTs have this potential. They can operate via many fields, for example *"in the management of urban systems or provide citizens with a more sustainable lifestyle" (Kramers et al., 2014).*

There are many studies about potential ICT solutions. The book written by William John Mitchell (2000) titled *"E-topia, Urban life, Jim-but not as we know it"* suggests five solutions focused on how ICT can contribute to the efficient reduction of energy in cities.

The first one is dematerialization; converting physical products into digital ones. The second is demobilization; all that has been digitized can be transported through via networks rather than being physically transported. The third is mass customization; less use of resources due to demand management and intelligent adaptation. Next is intelligent operation; preventing mismanagement and promoting more efficient management of resources. Last is soft transformation; transformation of physical infrastructure through the new opportunities that give us information.

Another point to take into account is the concept of **sustainability**. It is very important to relate this concept to new projects being implemented in cities to achieve the objectives of Strategy 2020. This field must take into consideration the **sustainable consumption** of both new technologies and the population. It is essential to understand that it is necessary to improve the efficiency of the technology but also follow some guidelines for sustainable consumption in society. An increased consumption by the citizens coinciding with a new found better efficiency of a technology is very common.

To achieve greater efficiency in the use of energy, the **rebound effect**[1] has to be taken into consideration *(Herring, 2004)*. The rebound effect, according to Dimitropoulos and Sorrell (2008), can be divided into three distinct categories:

- Due to a lower price of a product, consumers will spend more on the same products.
- Due to a reduction in the price of a product, consumers will spend more on other products, therefore consuming more energy.
- The effects of the economy caused by technological innovation and changes in consumer preferences lead to the creation of new products with low costs creating an increase in energy consumption.

A clear example of the rebound effect is that *"lighting today is 700 times more efficient than the lamps that were used in 1800. The effect is that the current consumption is over 6500 times greater than at that time"* (Herring et al., 2007).

1.3 Aims and objectives

This section provides the aims and objectives of the project.

1.3.1 Aims

The purpose of this thesis is to identify the role of sustainable consumption in smart sustainable cities' projects across Europe.

1.3.2 Objectives

To achieve this goal, the definitions of Smart Cities and Smart Sustainable Cities were reviewed and an overview of the Smart Cities projects in Europe was given. The report also analyzes a set of smart city projects in 76 cities in Europe classifying them by sectors

---

[1] The rebound effect is the term used to describe the effect that the lower costs of energy services, due to increased energy efficiency, has on consumer behavior both individually and nationally.

(smart governance, smart mobility, smart living, smart environment, smart citizens and smart economy).

In addition, the study examines the place of sustainable consumption in the projects evaluated. Finally, solutions to integrate sustainable consumption in projects are considered.

The report wants to achieve the following four objectives:

- To identify and analyze existing definitions of Smart Cities and Smart Sustainable Cities.
- To identify and select data sets about implementation of smart cities' projects in Europe for further analysis.
- To assess and analyze which level of sustainable consumption is relevant in the evaluated projects.
- To identify the best practices and provide the general recommendations for implementing sustainable consumption in smart cities.

2. Methodology

This thesis was conducted in three distinct phases over the spring semester, including a literature review, project analysis in the European Union and project evaluation of sustainable consumption.

2.1 Literature review

The initial part of the project consists of an extensive literature review of the cities' backgrounds and how their concept has been evolved. Apart from this, it is important to know how and why cities want to evolve and to achieve the aims of Europe 2020.

The research methodology included review papers, websites, databases of European projects and reports.

2.2 Analysis of projects in European Union

The second phase of the project is to analyze a set of projects carried out in some cities in the European Union. The projects analyzed were found in Concerto's database *(Dammann et al., 2005)*, the study of Mapping Smart Cities in Europe *(Manville et al., 2014)*, the report of Energy Study for the Stockholm Region *(Nylund, 2010)* and the report of Joint European Support for Sustainable Investment in City Areas *(Hirst et al., 2012)*.

The different sectors of smart cities, (smart governance, smart mobility, smart environment, smart living, smart citizens and smart economy) have led to the analyzed projects being classified into groups. In this way it is possible to find out the sectors in which the projects are more developed. Aside from that, two databases have been analyzed and compared, Concerto and Mapping Smart Cities in Europe. The smart city sectors' classification in the study of Mapping Smart Cities in Europe has already been carried out.

To define the different sectors of the smart city, the report has taken the following references: *(Neirotti et al., 2014), (Kramers et al. 2014), (Manville et al., 2014), (Giffinger, 2007), (Schurr, n.d.)*.

2.3 Evaluation of projects regarding sustainable consumption

Another important factor in the construction of this report was to consider if the projects included the concept of sustainable consumption or not. It will be attempted to discover if the stakeholders only think of the introduction of ICT solutions as more efficient or if they also take into consideration the sustainable consumption.

To analyze the concept of sustainable consumption, it is necessary to consider a number of parameters that classify projects. This can be split into four categories.

The references for that evaluation are *Sanne (2002), Banbury et al. (2012), Phipps et al. (2013) and Lorek et al. (2013)*. The categories are explained below:

> A: When the concept of sustainable consumption is not relevant in the project. In the planning and implementation of the project, there are no parameters referring to sustainable consumption. (SCo is not relevant).
>
> B: When the concept of sustainable consumption is relevant but it is not present in the project. Knowledge of sustainable consumption is present and clearly shows its position but it is not involved in the project. (SCo is relevant but not implemented)
>
> C: When the concept of sustainable consumption is relevant but only partly used in the project. The concept is very much present in the project but has not been fully exploited. (SCo is relevant and partly used).
>
> D: When the concept of sustainable consumption is relevant and is used to its full potential. The concept is clear and present and is crucial to achieving the objectives and is used in its totality. (SCo is relevant and used to its full potential).

The classification provides us the opportunity to estimate whether sustainable consumption is relevant or not in the projects evaluated. Finally, the report identifies a number of measures in projects to get closer to the higher criteria of sustainable consumption, and consequently, the cities may have more ambitious goals in the future.

3. Background

The background contains an explanation about the concept of Sustainable Consumption (SCo) and also a description of the different sectors of Smart Cities.

3.1 Sustainable Consumption

The concept of **Sustainable Development** was born through the Brundtland Report in 1987. Sustainable Development is the predominantly development that meets the needs of the present without compromising the ability of future generations to meet their own needs. There are two important concepts in this definition. The first one is the concept of needs; priority should be given to the world's poorest people and satisfy their basic needs. The second is the idea of limitations that are the environment's ability to meet both present and future needs *(Brundtland, 1987)*.

Sustainable Development requires special attention to two factors; the first is the rates of consumption of limited resources and the second is the available sink capacity for pollution. It is also necessary to include the distributional effects of consumption *(Carley et al., 1998)*.

An important concept that was born in the Rio Summit of the United Nations Conference on Environment and Development in 1992 was the **Agenda 21** action plan; a plan for future sustainability development. Agenda 21 offers ideas on how all levels of governance can take action to preserve natural resources, decrease the pollution and give some recommendations on sustainable development *(Steiner et al., 2013)*. Apart from that, Agenda 21 has a chapter called *"Changing Consumption Patterns"* which refers to *"new concepts of wealth and prosperity which allow higher standards of living through changed lifestyles that are less dependent on the Earth's finite resources and more in harmony with the Earth's carrying capacity"* *(Banbury et al., 2012)*. Agenda 21 declared that *"the major cause of the continued deterioration of the global environment is the unsustainable pattern of consumption and production, particularly in industrialised countries"* *(Carley et al., 1998)*.

The concept of **sustainable consumption** was also born in Rio 1992. There were 27 principles that were issued at the Summit. Principle eight refers to the *"Reduction of Unsustainable Patterns of Production and Consumption"* *(Banbury et al., 2012)*. In this principle, they made a link between sustainable development and sustainable consumption: *"To achieve sustainable development and a higher quality of life for all people, states should reduce and eliminate unsustainable patterns of production and consumption and promote appropriate demographic policies"*.

"There is actually no consensus on the concept of sustainable consumption" *(Neirotti et al., 2014)*. However, there is a definition of sustainable consumption created in 1994 by the Oslo Symposium on Sustainable Consumption made by the Norwegian government, NGOs and inter-governmental organizations. The definition is *"the use of goods and services that respond to basic needs and bring a better quality of life, while minimizing the use of natural resources, toxic materials and emissions of waste and pollutants over the life cycle, so as not to jeopardise the needs of future generations"*. In 1995, the Oslo Round Table on Sustainable Production and Consumption gave more clarity to the term: *"Sustainable consumption is an umbrella term that brings together a number of key issues, such as*

meeting needs, enhancing the quality of life, improving resource efficiency, increasing the use of renewal energy sources, minimising waste, taking a life cycle perspective and taking into account the equity dimension. Integrating these component parts is the central question of how to provide the same or better services to meet the basic requirements of life and the aspirations for improvement for both current and future generations, while continually reducing environmental damage and risks to human health. A key issue is therefore the extent to which necessary improvements in environmental quality can be achieved through the substitution of more efficient and less polluting goods and services (patterns of consumption), rather than through reductions in the volumes of goods and services consumed (levels of consumption). Political reality in democratic societies is such that it will be much easier to change consumption patterns than consumption volumes, although both issues need to be addressed."

In many countries, **energy efficiency** has become essential in the fight against Climate Change based on reduced consumption of energy and CO_2 *(Herring, 2004)*. It is necessary to reduce resources used and be more focused in developed countries where the resources consumed are much higher than in the developing countries. Furthermore, it is unclear whether consumption should be reduced or done differently and if individual consumers can contribute significantly to the conservation of resources *(Banbury et al., 2012)*. It is very important to emphasize that *"humanity uses 40% more resources in a year than nature can regenerate in the same year" (Williams, 2010)*.

Nowadays, it is possible to hear arguments such as *"it is absolutely necessary to improve energy efficiency, reduce energy consumption and thus reduce carbon emissions" (Herring et al.,2007)*. Moreover, many economists say a **rebound effect** occurs by increasing the efficiency of energy systems. This represents an increase in consumption due to the implicit price and makes it more affordable. *"Global energy demand grew 50% from 1980 to 2005, and this number is expected to increase another 50% by 2030" (Townsend, 2013)*.

Users raise their consumption without thinking about the effects this may have on the environment *(Herring, 2006)*. This effect varies depending on the cost and demand for energy services. A simple example of this would be that *"changing a 75 W light bulb with an 18 W bulb would reduce the energy used by 75%" (Herring et al., 2007)*. This is not happening due to the rebound effect. Therefore, the rebound effect may also vary between countries *(Herring, 2004)*.

While on the other hand, there are other approaches to reduce consumption behaviour. Not just the price influences consumption but also there are social factors that are also open *(Manoochehri, 2002)*. If society wants to have a more sustainable standard of living, it is absolutely necessary for consumers to modify the goods consumed and reduce consumption levels *(Buenstorf, 2008)*. Citizens are stakeholders that directly influence the consumption of goods and services and must make a difference to the environment *(Spaargaren, 2008)*. It is necessary to clarify that *"sustainable consumption is not only for producing goods with less energy, it is also the lifestyle of people; where and how we live, what we eat, etc." (Lorek et al., 2013)*.

Humans have an increasing need to consume and it is not provable that the reduction of consumption will be voluntary. This is also induced by an increase in productivity and technological innovation which we are bombarded with daily in the form of new consumption opportunities *(Buenstorf et al., 2008)*. If we want to pursue a sustainable state, we have to make a change in consumers. They will have to reduce their consumption and change the way that they consume goods and services *(Buenstorf et al., 2008)*.

The paper written by Lorek and Fuchs in 2013 distinguishes between two different types of sustainable consumption. The first is the strong sustainable consumption and the second is the weak sustainable consumption. Strong sustainable consumption emphasizes the need for a change in the levels of consumption by citizens. They make consumption decisions. The weak sustainable consumption refers the low consumption that can be achieved through efficiency due to the benefits of new technologies. That means production in the most efficient way.

3.2 Sectors of Smart Cities

The projects related to smart cities can be classified into different sectors according to the objectives they present. Projects may be involved in different aspects to improve the functions of the city. Some examples would be the enhancement of quality of life for citizens, the reduction of energy consumption and CO_2 emissions, effective and efficient transport, etc.

It is important to know which type of project is in each sector. Furthermore, it is necessary to clarify the different sectors that interact with each other in Smart Cities. After that, it is possible to classify which kind of project is in each sector.

To define the different sectors of the smart city, the report has taken the following references: *(Neirotti et al., 2014), (Kramers et al., 2014), (Manville et al., 2014), (Giffinger, 2007), (Schurr, n.d.)*.

3.2.1 Smart Economy

"Smart Economy means e-business and e-commerce" *(Manville et al., 2014)*. Companies that wish to bet on a smart economy should have a close relationship with ICT. From developing ICTs, new products and services are created *(Neirotti et al., 2014)*. Companies must be connected to each other, sharing knowledge and getting an overview of the local and global economy *(Manville et al., 2014)*. Furthermore, *"it is important to foster the innovation systems and entrepreneurship in the urban ecosystem"* *(Neirotti et al., 2014)*.

1. *"Innovative spirit and entrepreneurship" (Giffinger, 2007 and Neirotti et al., 2014).*
   Companies must have an innovative character and a high level of entrepreneurship.

2. *"Productivity" (Giffinger, 2007).*
   Companies should have smart production rates to operate in an efficient manner and increase profits. They should look into how they can increase their production with fewer resources.

3. *"Ability to transform" (Giffinger, 2007).*
   It is necessary that companies should have the ability to adapt their production to the need of the population. They have to improve the products when necessary.

3.2.2 Smart Mobility

'Smart Mobility' means an interconnection net between the subway, buses, trains, trams, cars, bicycles and citizens *(Manville et al., 2014)*. They should all be able to coordinate their lives in the most efficient way. People are in constant transit and so transport should be a dominating factor in a Smart City *(Manville et al., 2014)*. This is a difficult task and companies and governments should work together to achieve efficiency and usefulness for the citizens through the following means;

1. *"To build the necessary infrastructure through demand forecasting" (Schurr, n.d.)*.
   Accurate prediction of demand is absolutely necessary before development of the infrastructure. Prediction techniques should incorporate sensors and GPS data so that time is not wasted on displacements *(Schurr, n.d.)*.

2. *"To calculate the best offset from its origin to its destination" (Shurr, n.d.)*.
   Citizens can use tools to optimize their travel and they can find the time, cost and environmental impacts arising from these. Cities that have inefficient services generate excess costs and high environmental impacts for the citizens *(Schurr, n.d.)*.

3. *"Assure safety and security" (Schurr, n.d.)*.
   In the field of smart mobility, it is important to consider the concepts of safety and security. The issues and risks arising from testing and using sensors and cameras should be discussed.

3.2.3 Smart Environment

'Smart Environment' looks at a reduced dependence on fossil fuels and an introduction of renewable energies in the system. The CO_2 emissions should be reduced so that people can live in a healthier environment. As well as this, remodelling the city with projects that reduce energy consumption in buildings and lighting in the city are addressed. Aside from that, cities would also be involved in projects about waste management, water resources and reducing pollution *(Neirotti et al., 2014)*.

1. *"Smart grids and public lighting" (Neirotti et al., 2014)*.
   Smart grids use a number of tools for citizens to calculate the energy they consume. The grids are also helpful with regard to the services used for an effective energy management distribution *(Neirotti et al., 2014 and Schurr, n.d.)*.
   The street lights consume a great amount in the city and must be managed properly. If management is done right, it can reduce maintenance costs, energy and CO_2 emissions *(Neirotti et al., 2014)*.

2. *"Renewable energies" (Neirotti et al., 2014)*.
   It is necessary to reduce dependence on fossil fuels and introduce renewable energy through natural resources *(Neirotti et al., 2014)*.

3. *"Waste and water management" (Neirotti et al., 2014 and Schurr, n.d.)*.
   When thinking Smart Environment, it is absolutely necessary effectively manage waste and water. Regarding waste, it is very important to properly assess classification of waste and manage them properly. Moreover, it is important to

manage water resources well and carry out analysis of the quality and quantity that is needed for citizens *(Neirotti et al., 2014)*.

4. "Food and agriculture" *(Neirotti et al., 2014)*.
   The technology can also affect the field of agriculture. It has the ability to regulate crop conditions via moisture, light and temperature sensors *(Neirotti et al., 2014)*.

3.2.4 Smart Society

'Smart Society' incorporates creative people with innovative ideas. The citizens are involved in city projects and make their own decisions about where they want to live. They have the capacity to improve their skills and have an affinity for lifelong learning. They are also able to create products and services. The citizens who live in a smart city must be flexible and educated *(Manville et al., 2014)*.

1. *"Level of qualification" (Giffinger, 2007).*
   A smart society is characterized by a high level of qualification.

2. *"Affinity to lifelong learning" (Giffinger, 2007).*
   This society has to be ready to learn everyday about the different sectors that the city is involved in and understand the guidelines that their city goes by.

3. *"Flexibility and creativity" (Giffinger, 2007).*
   They should be creative and flexible to adapt to different situations.

4. *"Participation in public life" (Giffinger, 2007).*
   The citizens can participate in projects taking place in the city. They should have high participation levels and be impartial when making decisions.

3.2.5 Smart Living

'Smart Living' addresses the quality of life people have in the city. Furthermore, these citizens live in a safe and healthy environment in good houses with healthy facilities that a Smart City gives them. They have good access to cultural, health and education institutions that make life easier and more comfortable *(Manville et al., 2014 and Neirotti et al., 2014)*.

1. *"Cultural and education facilities" (Giffinger, 2007).*
   The city should convey information about cultural activities to the citizens to enjoy in their free time and depict a wide knowledge of different areas. Also, it is important to motivate people to get involved and participate in them *(Manville et al., 2014)*. Public schools are provided with modern ICT tools *(Neirotti et al., 2014)*. The infrastructures should generally be close to all citizens, creating a local and communal feel.

2. *"Health care" (Neirotti et al., 2014).*
   Citizens will feel safer when medical facilities providing early diagnosis and treatments for all medical needs *(Neirotti et al., 2014)*.

3. *"Public safety" (Giffinger, 2007 and Neirotti et al., 2014).*

Local public organization should protect citizens and their possessions in the cities they live in so that citizens can feel safe in their communities *(Neirotti et al., 2014)*.

4. *"Housing quality" (Giffinger, 2007 and Neirotti et al., 2014).*
   It is very important to provide citizens with a high comfort level in buildings. Heating, lighting, ventilation, etc. are priorities in the provision of quality housing *(Neirotti et al., 2014 and Kramers et al., 2014)*.

5. *"Social Cohesion" (Giffinger, 2007).*
   The city is characterized by a big network of citizens where social cohesion comes into focus.

3.2.6 Smart Governance

The Smart Governance concept refers to the security of public services that use technology to facilitate and support better planning and decision making. It is important to make the decisions in the most democratic form so that the entire population can have the opportunity to exercise their vote *(Manville et al., 2014)*.

1. *"E-government" (Neirotti et al., 2014).*
   This tool allows citizens to use newer and faster services through the digitalization of public administration *(Neirotti et al., 2014)*.

2. *"Transparent governance" (Giffinger, 2007).*
   The citizens can access in a simple manner official documents for participation in the decision processes of a municipality.

3. *"E-democracy" (Neirotti et al., 2014).*
   ICT's are used to improve citizen participation for policies in a democratic manner. *(Neirotti et al., 2014)*.

4. Results

The results are divided into four sections. The first and the second section are about the identification and analysis of the definitions of smart city and smart sustainable cities, where the commonalities and differences are identified across different definitions. The third section is about the analysis of smart sectors and the last section is about the relevance of sustainable consumption in projects evaluated.

4.1 Definitions of Smart Cities

There are lots of definitions about Smart Cities and they come in many variations, sizes and types. The idea of a Smart City is relatively new and for this reason it is very broad. Every city has its own history, current characteristics and future possibilities that make it unique.

If we take a look at the differing definitions of Smart Cities, the concept is a mix of technologies, social and economic factors and policy and business drives.

There are various definitions of Smart Cities detailed below:

1. *"Smart City, in everyday use, is inclusive of terms such as digital city or connected cities. Smart Cities as an applied technology term often refers to smart grids, smart meters, and other infrastructure for electricity, water supply, waste and refers to 'city basic".*

   (Hire, Christopher, Innovation cities programme).

If we take a look to the definition above, it means that Smart Cities are strongly linked with technology. The Smart City attempts to use the technology to make life easier for the citizens.

2. *"A Smart City is a well performing city built on the 'smart' combination of endowments and activities of self-decisive, independent and aware citizens."*

   (Giffinger et al., 2007: 11)

This definition shows that the citizens should be involved with a project and should participate in the city through smart actions. The only concept that the author considers is regarding the ordinary people. He does not give mention to the government, natural resources or technology.

3. *"In a Smart City, networks are linked together, supporting and positively feeding off each other, so that the technology and data gathering should: be able to constantly gather, analyse and distribute data about the city to optimise efficiency and effectiveness in the pursuit of competitiveness and sustainability; be able to communicate and share such data and information around the city using common definitions and standards so it can be easily re-used; be able to act multi-functionally, which means they should provide solutions to multiple problems from a holistic city perspective".*

(Copenhagen Cleantech Cluster, 2012)

This definition of a SC demonstrates the use of technology as the key to achieving its objectives such as analysis and distribution of data to optimize efficiency and effectiveness in pursuit of ultimate sustainability. The city must be able to have solutions to the problems that arise.

4. *"A smart city is where the ICT strengthens freedom of speech and the accessibility to public information and services".*

(Anthopoulos and Fitsilis, 2010)

This definition only highlights the work of ICT to provide information to citizens and have access to public services.

5. *"Smart Cities are about leveraging interoperability within and across policy domains of the city (e.g. transportation, public safety, energy, education, healthcare, and development). Smart City strategies require innovative ways of interacting with stakeholders, managing resources, and providing services".*

(Nam and Pardo, 2011)

Nam and Prado declare a SC as one that connects different sectors of the city, i.e. there is an interaction between resource management and the provision of services.

6. *"Smart Cities combine diverse technologies to reduce their environmental impact and offer citizens better lives. This is not, however, simply a technical challenge. Organisational change in governments – and indeed society at large – is just as essential. Making a city smart is therefore a very multi-disciplinary challenge, bringing together city officials, innovative suppliers, national and EU policymakers, academics and civil society".*

(Smart Cities and Communities, 2013)

This definition has prioritized the use of new technologies to provide citizens with a better life, and concurrently, reduce the environmental impacts of the city. Furthermore, to create a SC, innovative people, government, researchers and others are required to work together to achieve the goal.

7. *"A city may be called 'smart' when investments in human and social capital and traditional and modern communication infrastructure fuel sustainable economic growth and a high quality of life, with a wise management of natural resources, through participatory governance".*

(Schaffers et al., 2011)

In the definition of a SC described by Schaffers, several factors are highlighted. These are high quality of life, sustainable economic growth, investment in human and social capital, etc. which are all achieved through citizen participation.

8. *"Smart City is a city that uses data, information and communication technologies strategically to provide efficient service to citizens, monitor policy outcomes,*

manage and optimise existing infrastructure, employ cross-sector collaboration and enable new business models".

(The Climatic Group et al., 2011)

The definition provided by The Climate Group gives more importance to the new technologies providing a service to citizens by creating a network where everything is connected and new business models based on technology will consequently be created.

As mentioned above, it is possible to see different definitions according to the author, but they all have some commonalities such as sustainable economic growth, proper management of resources, citizen interaction and technology. The latter, technology, is present in most of the definitions. It plays a big role in improving the skills of a city and must be present in the projects. Nevertheless, most definitions of SC give very little importance to the environmental performance of cities.

4.2 Definitions of Smart Sustainable Cities

The concept of a Smart Sustainable City is relatively new and it is not overly clear. It is possible to find different definitions about the term from companies, research groups and governments. Cities that are committed to be smart sustainable cities have very ambitious goals and have to work hard to achieve them. These cities want to achieve the objectives with the help of new technologies.

There are some definitions about the term of smart sustainable cities below:

1. *"A smart and sustainable city invests in human and social capital, manages resources wisely, has citizens which participate in city governance, and has traditional and modern infrastructure which supports economic growth to create high quality of life for its inhabitants".*

   (JESSICA: Joint European Suport for Sustainable Investment in City Areas, 2012)

In the definition of a SSC created by JESSICA, there is referral to concepts such as; investment in human and social capital, a high quality of life for citizens, high citizen participation to work with the government on projects and decisions of the city, support for economic growth but without the concept of sustainability and promotes resource management satisfactorily.

2. *"A Smart Sustainable City is settlements where investments in human and social capital and traditional (transport) and modern (ICT) communication infrastructure fuel sustainable economic growth and a high quality of life, with a wise management of natural resources, through participatory governance".*

   (Caragliu, A., Del Bo, C and Nijkamp, P. (2011) Smart Cities in Europe, Journal of Urban Technology, vol. 18, (2): 65-82).

This definition concentrates on participatory governance to achieve a SSC, working on the concepts of sustainable economic growth, high quality of life, rational management of natural resources, technology, etc.

3. *"Smart Sustainable Cities use information and communication technologies (ICT) to be more intelligent and efficient in the use of resources, resulting in cost and energy savings, improved service delivery and quality of life, and reduced environmental footprint--all supporting innovation and the low-carbon economy".*

    (Cohen, Boyd, 2011)

This definition of Smart sustainable cities proposes that the use of new technologies would be mandatory in order to reduce energy costs and increase the quality of life and the delivery of services. It also refers to reducing the carbon footprint throughout sustainable growth.

4. *"The Smart Sustainable City seeks to achieve concern for the global environment and lifestyle safety and convenience through the coordination of infrastructure. Smart Sustainable Cities realized through the coordination of infrastructures consist of two infrastructure layers that support consumers' lifestyles together with the urban management infrastructure that links these together using IT".*

    (Hitachi, 2013)

In the definition made by Hitachi there is clear responsibility for environmental concerns Citizens have a high quality of life, adequate safety standards, and also a good service management. New technologies are used as a tool to achieve the objectives giving its SSC status.

5. *"A smart sustainable city is a city that leverages the ICT infrastructure in an adaptable, reliable, scalable, accessible, secure, safe and resilient manner in order to:*
    - *Improve the Quality of Life of its Citizens.*
    - *Ensure tangible economic growth such as higher standards of living and employment opportunities for its citizens.*
    - *Improve the well-being of its citizens including medical care, welfare, physical safety and education*
    - *Establish an environmentally responsible and sustainable approach which "meets the needs of today without sacrificing the needs of future generations".*
    - *Streamline physical infrastructure based services such as the transportation (mobility), water, utilities (energy), telecommunications, and manufacturing sectors.*
    - *Reinforce prevention and handling functionality for natural and man-made disasters including the ability to address the impacts of climate change.*
    - *Provide an effective and well balanced regulatory, compliance and governance mechanisms with appropriate and equitable policies and processes in a standardized manner".*

    (ITU-T Focus Group (WG 1) on Smart Sustainable Cities, 2014)

The definition above has prioritized the use of new technologies to achieve an improvement in the quality of life for citizens in every way. The environment plays an important role which should be preserved and it also demonstrates the ability to fight the impacts of climate change. Furthermore, it is also important that the compliance and governance mechanisms, with appropriate management, provide good policies and processes in a standardized manner.

6. *"A Smart Sustainable City is one in which the seams and structures of the various urban systems are made clear, simple, responsive and even malleable via contemporary technology and design. Citizens are not only engaged and informed in the relationship between their activities, their neighbourhoods, and the wider urban ecosystems, but are actively encouraged to see the city itself as something they can collectively tune, such that it is efficient, interactive, engaging, adaptive and flexible, as opposed to the inflexible, mono-functional and monolithic structures of many 20th century cities".*

    (Hall et al., 2009)

Here, technology is interpreted as a possible solution to intervene in urban systems and improve their services. An important factor to take into account is the citizens who are present in the projects carried out in the city. They must decide on the future of this. Cities have to be malleable to transform and improve upon the request of citizens.

7. *"We define Smart Sustainable City as the city that uses information technology and communications to make both their critical infrastructure, its components and utilities offered more interactive, efficient and citizens to be more aware of them. It is a city committed to the environment, both environmentally and in terms of cultural and historical elements".*

    (Telefonica, n.d.)

The definition given by Telefonica relates to the use of technology to improve city services and the efficiency of resources. It also refers to the smart sustainable city concept as a city committed to the environment while still keeping in mind the history of cities.

8. *"Smart Sustainable Cities combine diverse technologies to reduce their environmental impact and offer citizens better lives. This is not, however, simply a technical challenge. Organizational change in governments - and indeed society at large - is just as essential. Making a city smart is therefore a very multidisciplinary challenge, bringing together city officials, innovative suppliers, national and EU policymakers, academics and civil society".*

    (European Commission, n.d.)

The key idea from this definition relate to reducing the environmental impact of cities and citizens gaining a better lifestyle through technology. The transformation of the city is a multidisciplinary challenge in which a large part of society acts in the best interests of the city.

9. *"A Smart Sustainable City is mainly based on the information and communication technologies. Through the transparent and full access to information, the extensive and secure transmission of information, the efficient and scientific utilization of*

information, it increases the urban operational and administrative efficiency, improves the urban public service level, form the low-carbon urban ecological circle, and construct a new formation of urban development".

(ITU Focus Group on Smart Sustainable Cities, 2013)

In this definition, technology is used as the primary matter to achieve the requirements of SSC. Through the achievements of new technologies, urban development is built in a sustainable and efficient way.

10. *"Smart Sustainable Cities are well managed, integrated physical and digital infrastructures that provide optimal services in a reliable, cost effective, and sustainable manner while maintaining and improving the quality of life for its citizens. Key attributes of a Smart Sustainable City are Mobility, Sustainability, Security, Reliability, Flexibility, Technology, Interoperability and Scalability. Foundational aspects include Economy, Governance, Society and Environment with Vertical Infrastructures such as Mobility, Real Estate & Buildings, Industrial & Manufacturing, Utilities-Electricity & Gas, Waste, Water & Air Management, Safety & Security, Healthcare and Education. All of these are woven into a single fabric with ICT infrastructure as a core".*

(Kondepudi, 2013)

The definition from Kondepudi (2013) refers to SSC as places that provide a high quality of life for citizens by offering them quality services in a reliable and sustainable way. It also defines the key points that the city has to have such as mobility, security and flexibility to achieve the objectives through new technologies.

11. *"A Smart Sustainable City has been defined as a 'knowledge', 'digital', and 'cyber' or 'eco' city; representing a concept open to a variety of interpretations, depending on the goals set out by a Smart Sustainable City's planners. We might refer to a Smart Sustainable City as an improvement on today's city both functionally and structurally, using information and communication technology (ICT) as an infrastructure. Looking at its functions as well as its purposes, a Smart Sustainable City can perhaps be defined as "a city that strategically utilizes many smart factors such as Information and Communication Technology to increase the city's sustainable growth and strengthen city functions, while guaranteeing citizens' happiness and wellness."*

(Hwang et al., 2013)

According to this research group, they define a SSC as an open concept that varies with the objectives of the planners. The use of technology for sustainable growth of the city is necessary and also it strength its functions.

In many of the definitions that have been looked at, the role of new technologies is very important in achieving the Sustainable City status. It is possible to see some commonality with the definitions of smart cities and smart sustainable cities. However, in the definitions of smart sustainable cities it can be said that they give more importance to the environmental impacts, sustainability, resource efficiency, cost and energy savings, environmental responsibility, etc.

Projects undertaken in the different cities will be looked at with regard to the definitions examined. The projects will have to show sufficient promise to achieve the goals that the city needs in order to be classified as a SSC.

4.3 Analysis of the smart sectors

To analyze the different projects evaluated, 148 projects in Europe have been classified. Projects were assessed according to different sectors. The references for the classification were taken from *Neirotti et al. (2014), Kramers et al. (2014), Manville et al. (2014), Giffinger (2007) and Schurr (n.d.)*.

The projects' classifications of the different sectors and the sustainable consumption are detailed in Appendix I of this report. Fragment of appendix 1 is given in to provide an example.

Country	City	Smart solutions technology	S. Governance	S. Mobility	S. Living	S. Environment	S. Society	S. Economy	S. Consumption	Explanation of sustainable consumption grade
France	Ajaccio	Solar ventilation system, photovoltaic panels.			X	X			C	They use the right technology but they do not implemented patterns to reduce consumption of users.
Denmark	Copenhagen	Intelligent traffic system, monitoring and sensors, e-bikes.		X		X			D	Strong prioritizing of bicycle-friendly infrastructure, consciousness of people to use the bicycle and public transport instead of private transport.
Austria	Vienna	Urban information models, data smart meters, mobile devices.		X	x	X	x		C	The project aims the optimization of energy consumption, improving inclusion and mobility of people through the technology.
Sweden	Malmo	Open data, apps, e-skills.	X					X	A	The project helps to use open data to create mobile applications to live more intelligently.

The Netherlands	Amsterdam	Public available WiFi	X	x		x	B	The project highlight is to have a greater connection services. The project thinks about the sustainable consumption but it is not present on it.

Table 1: Fragment of the projects' classification to illustrate how it was organized

Appendix I provides us the classifications of the smart cities' sectors and also the different categories about sustainable consumption. The classification has been done based on the information of the project that can be found online. The parameters for the classification of the categories for the sectors of smart cities and sustainable consumption are explained in chapter 3.2; sectors of smart cities, and in chapter 2.3; methodology.

After the classification of 148 projects, it is possible to analyze which kinds of smart city sectors are present in them.

The graphic below shows us the classification of the different sectors in the projects evaluated.

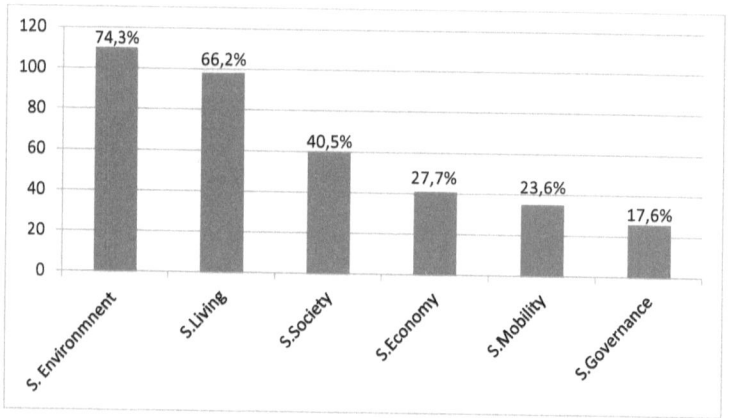

Figure 3: Classification of the different sectors in the projects evaluated

As we can see in Figure 3, Smart Environment and Smart Living is present in over 50% of projects; Smart Environment (74,3%) and Smart Living (66,2%). Smart Society (40,5%) and Smart Economy (27,7%) are the third and fourth respectively, and Smart Mobility (23,6%) and Smart Governance (17,6%) are fifth and sixth respectively.

It is possible to see that most of the projects evaluated are in the categories of Smart Environment and Smart Living because they integrate innovative energy efficiency measures with a substantial contribution from decentralised renewable energy sources, smart grids, renewable based cogeneration, district heating/cooling systems and energy management systems in buildings.

A comparison has been made from Concerto and Mapping Smart Cities in Europe databases. The next figure shows us the differences and communalities about these two datasets.

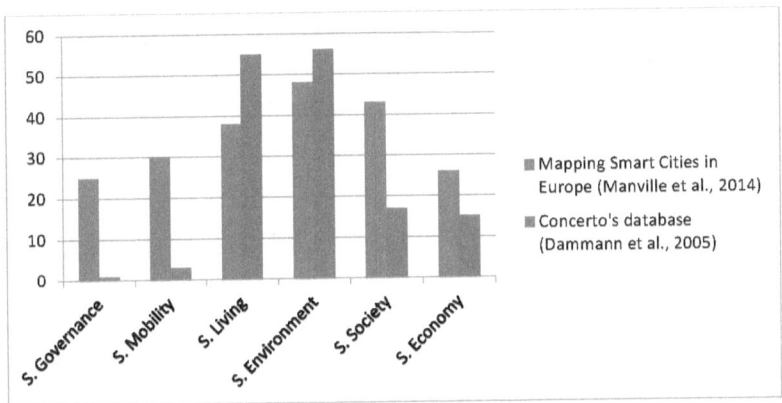

Figure 4: Comparison between Concerto's database and Mapping Smart Cities in Europe

In Figure 4, it is possible to appreciate that both datasets have the same communality; most of the projects are in the sector of Smart Environment and both have a less number of projects in Smart Governance. In the other hand, the Mapping Smart Cities in Europe database has a more regular distribution of projects while most of the projects in Concerto's database are in the sectors of Smart Environment and Smart Living, and the quantity of projects involved in Smart Governance and Smart Mobility is low.

The next figures show us the countries in which the sector of smart cities are located and also the percentage of projects per country in each sector. The first one is about the projects of Smart Governance (Figure 5); the second is about the projects of Smart Mobility (Figure 6); the third is about the projects of Smart Living (Figure 7); the fourth is about the Smart Environmental projects (Figure 8); the fifth is about the Smart Society projects (Figure 9) and the last one is about projects of Smart Economy (Figure 10).

Figure 5 represents countries in which there are projects about Smart Governance, and the corresponding number of cities.

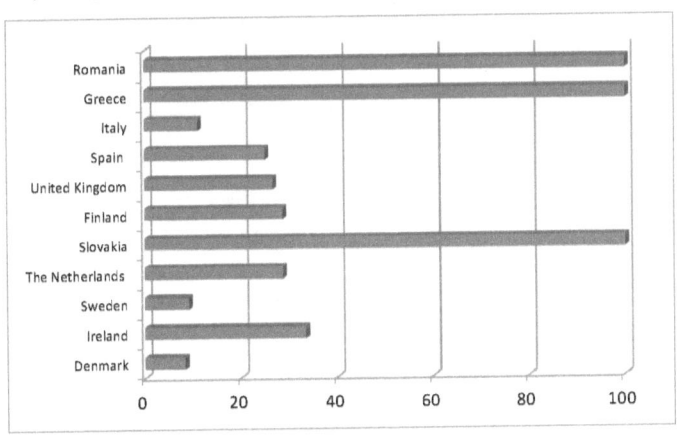

Figure 5: Percentage of Smart Governance projects in Europe

According to Figure 5, the SG sector, Romania and Slovakia can be ignored as they do not have a large enough sample having only one project evaluated. Greece contains this sector integrated in all the projects evaluated (3) and the countries with the next largest sample are Ireland, The Netherlands, Finland, United Kingdom and Spain. The projects that are involved in this sector are about apps, 4g wireless network, cloud computing, future of the public service, etc. The technology is used to facilitate and support better planning and decision making.

Figure 6 shows us the projects about Smart Mobility evaluated in Europe.

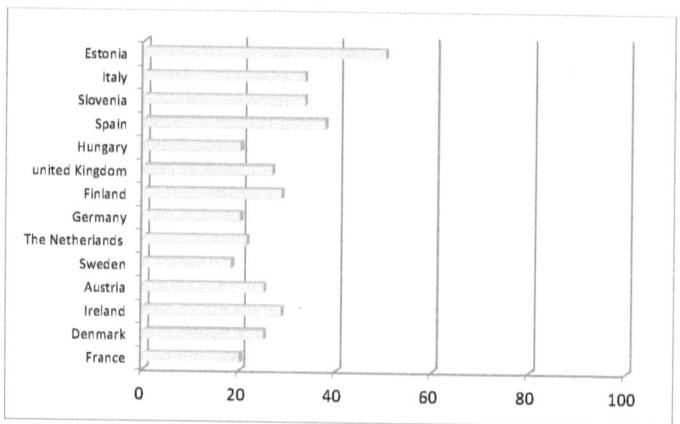

Figure 6: Percentage of Smart Mobility projects in Europe

According to Figure 6, the SM sector, we can appreciate that most of the countries have between twenty and forty percent of projects involved relating to Smart Mobility. Estonia is the only country that has more than 40% of its projects related to this sector. When these kinds of projects are analysed in depth, it can be seen that some of them are capital cities. They also all have a population greater than 500,000 apart from Ljubljana, Weiz and Tallinn. Projects involved in that sector are about car-sharing, electric vehicles, transport infrastructures, promotion of the bike and public transport, network traffic management, etc.

Figure 7 shows us the projects involved in Smart Living.

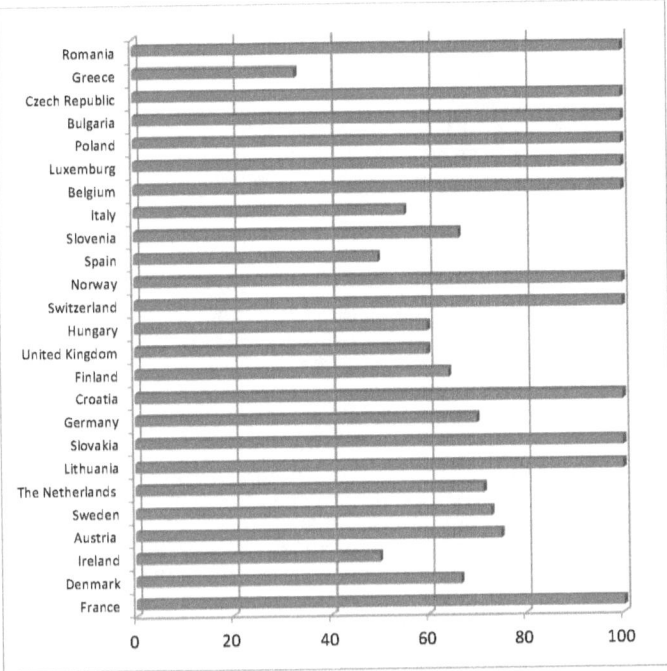

Figure 7: Percentage of Smart Living projects in Europe

As we can appreciate from Figure 3, there is a huge number of countries that are working in the Smart Living parameters.

According to Figure 7, the SL sector, it is important to note that almost all countries have over 50% of projects considered in this sector except Ireland (42.8%) and Greece (33.3%). The projects that are in this sector are involved in contributing to a better quality of life with regard to comfort, heating and lighting of the residential buildings, etc. Also there are some projects that provide public schools with an extensive use of ICT tools and provide a good atmosphere in which to live in.

To build and develop SSC, it is necessary to adapt new technologies for them. Buildings consume a lot of energy and contribute a lot toward the overall impact of cities. For this reason, most of the projects in this sector are considering how to reduce the consumption of the inhabitants and save energy while improving the infrastructure. It is possible to find a huge number of projects involved in this sector.

In Figure 8 it is possible to see the projects that involve Smart Environment in Europe.

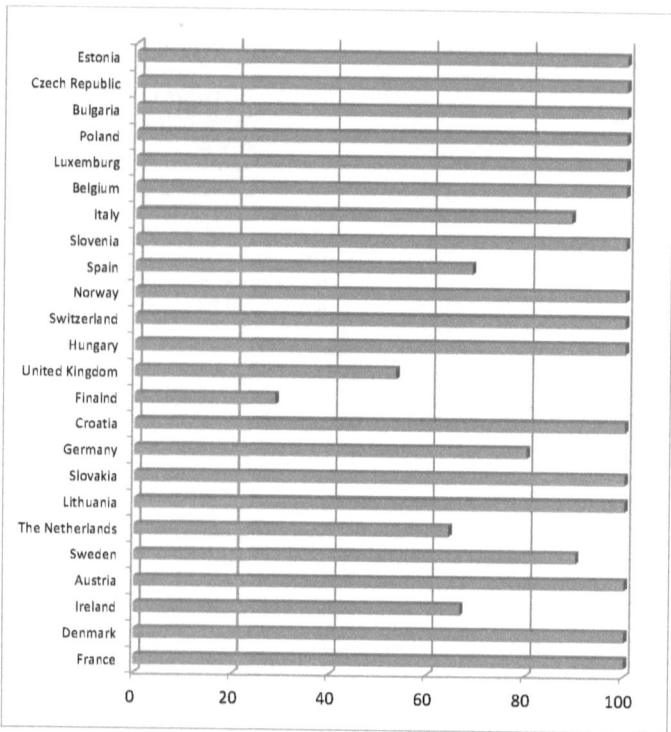

Figure 8: Percentage of Smart Environment projects in Europe

According to Figure 8, there is a large number of projects involved in SE. All of them are over 50% except Finland. The countries such as Italy, Spain, United Kingdom, Germany, The Netherlands, Sweden and Ireland have a lower percentage as within the set of projects evaluated, some of them have projects involved in other fields like mobile devices, 4g wireless network, real time traffic information, etc, that are not involved in the Smart Environment sector. Moreover, in other projects, it is possible to find some ICT solutions such as solar panels, biogas production for heat and electricity, wind energy, sorting and recycling, low energy lighting, etc.

It is remarkable that most of the projects in this sector are concerned with reducing CO_2 emissions and the energy consumption of citizens. It is a big challenge for cities to provide a good atmosphere, reducing their consumption and provide the population with all of their needs.

Figure 9 shows us the projects involved in Smart Society sector in Europe.

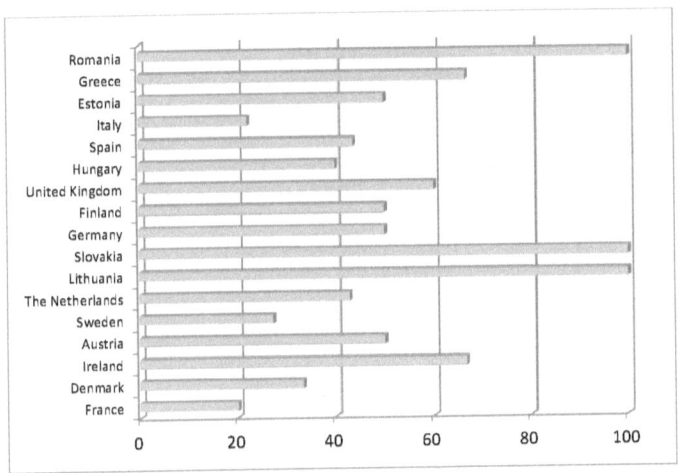

Figure 9: Percentage of Smart Society projects in Europe

According to Figure 9, it is possible to analyze the Smart Society's sector. It represents the third most common in the projects evaluated (40.5%). Romania, Slovakia and Lithuania can be ignored as they have only one project to choose from. Countries such as Greece, Estonia, United Kingdom, Finland, Germany, Austria and Ireland have 50% or more projects in this sector.

If a society is considering transforming their city into a smart sustainable city, it is entirely necessary to provide citizens with information about how to live in it and what they can do to achieve the aims of the city. SC requires citizens with a high level of education and they should also have an affinity for lifelong learning. For this reason it is not strange to find a large quantity of projects involved in this sector.

Figure 10 shows us the projects about Smart Economy.

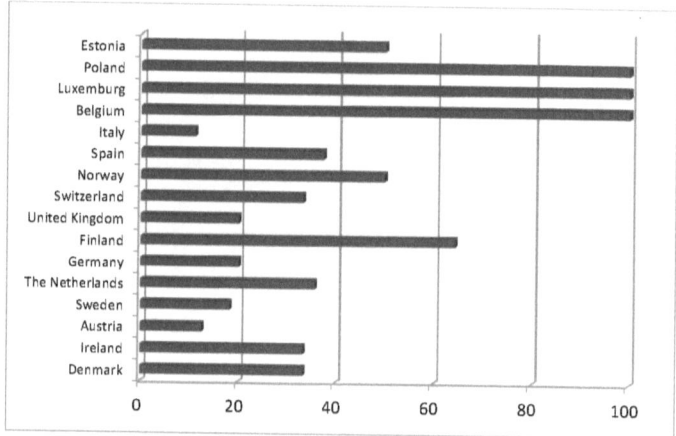

Figure 10: Percentage of Smart Economy projects in Europe

According to Figure 10, the SEc sector, Poland, Luxemburg and Belgium can be overlooked as there is only one project evaluated in each country. As we can see from Figure 10, countries that have the most projects involved in this sector are Finland, Norway and Estonia.

To achieve a smart sustainable city, it is necessary to have sustainable economic growth. Most of the projects involved in this sector look at creating new jobs, smart production to produce in an efficient manner, ability to adapt the production to the need of the population, etc. When analyzing the countries that have more projects in this sector, it can be seen that they are northern European countries.

4.4 Relevance of sustainable consumption

The references for the different categories of sustainable consumption evaluated are *Sanne (2002), Banbury et al. (2012), Phipps et al. (2013), Lorek et al. (2013)*. Projects have been separated into four categories according to the weight that sustainable consumption has in each project; A (SCo is not relevant), B (SCo is relevant but not implemented), C (SCo is relevant but partly used) and D (SCo is relevant and used to its full potential).

After the evaluations of the different sectors of smart cities, it is possible to predict the percentage of the projects' categories of sustainable consumption for each sector of the smart city. It has been convenient to separate Category A from the others because it is not representative of sustainable consumption projects. In short, Category A has various projects where the sustainable consumption concept is not relevant. This category has a total of 28 projects of the 148, which represents 18,9% of all the projects evaluated.

The percentage of Category A is for the total number of projects in all sustainable consumption categories (A, B, C and D) and the percentage of B, C and D categories is only for these categories. Therefore, the sum of all categories is not 100%, only the sum of the categories B, C and D is 100%.

	% of projects	Smart Governance	Smart Mobility	Smart Living	Smart Environment	Smart Society	Smart Economy
SCo priority ranking		6	4	2	1	3	5
A: SCo is not relevant	18,9%	42%	32%	14%	1%	20%	44%
B: SCo is relevant but not implemented	8,3%	53%	4%	6%	1%	15%	4%
C: SCo is relevant and partly used	54,2%	27%	61%	50%	58%	52%	48%
D: SCo is relevant and used to its full potential	37,5%	20%	35%	44%	41%	33%	48%

Table 2: Percentage of sustainable consumption categories per sector of smart cities' projects evaluated

In Table 2 we can see the percentage of sustainable consumption in all the sectors of smart cities. After analyzing the Smart Governance sector, it can be confirmed that this sector has very little projects involved in sustainable consumption. The greatest amount of these projects is in Categories A and B. Category A represents 42% with Category B representing 53%. The other categories of sustainable consumption are C with 27% and D with 20%. Most of the projects involved in Smart Governance are not related to sustainable consumption.

Category A considers the Smart Mobility sector representing 32% of the projects. The explanation is that these kinds of projects contain the real time of cities' buses, digital urban services that make travelling and living in the city easier, etc. On the other hand, there are a lot of projects in Categories C and D, 61% and 35% respectively. Category B represents 4%. The projects involved in the smart mobility sector follow the guidelines of sustainable consumption.

Regarding the sector of Smart Living, we can appreciate that Category A contains 14% of the projects, but almost 95% of projects in Categories C and D, 50% and 44% respectively. Category B has just 6% of the projects. If we analyze all the projects involved in this sector it is possible to affirm that most of them consider sustainable consumption (94.1%).

The Smart Environment sector has the most projects following the guidelines of sustainable consumption. There is only one project in both Categories A and B of sustainable consumption that represent 1% of each category. Category C represents 58% of the projects and Category D has 41% of projects evaluated. After this analysis it is possible to affirm that almost all of the projects evaluated in this sector follow guidelines of sustainable consumption.

Projects involved in the Smart Society sector contain 20% of the projects in Category A. These kinds of projects are about internet services, traffic information, mobile devices, etc. Category B contains 15% of projects, Category C contains 52% of and in Category D, 33% of the projects are evaluated. To build a smart and sustainable city, both full collaboration of smart citizens and their involvement in the projects is necessary. Smart societies have to take advantage of the possibilities that cities can give them. This sector is following guidelines of sustainable consumption patterns.

Going more into depth into the analysis of sustainable consumption of the Smart Economy sector, it can be seen that most projects are involved in Category A of sustainable consumption (44%). If Category A is discounted, it can be seen that the Smart Economy sector is related to Categories C and D of sustainable consumption. Both categories hold the same percentage of projects (48%). Therefore it can be said that projects related to Smart Economy are moving towards sustainable consumption patterns.

Looking at the classification in Table 2 of the sectors of smart cities and the relevance of sustainable consumption in the projects evaluated, it is possible to establish a ranking. The classification from lower to higher of the relevance of sustainable consumption reads Smart Governance, Smart Economy, Smart Mobility, Smart Society, Smart Living and Smart Environment.

The figure below shows us the percentage of relevancy of sustainable consumption for all the projects evaluated. The second graph shows us only the percentage of the projects evaluated in Categories B, C and D, where SCo is relevant.

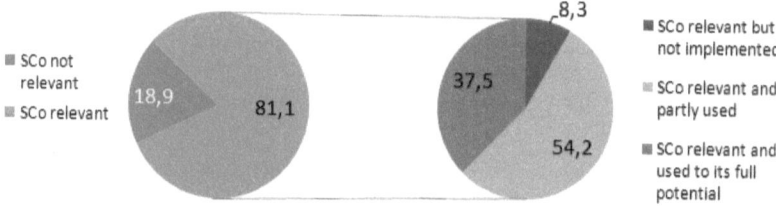

Figure 11: Percentage of the projects evaluated in sustainable consumption categories

As we can appreciate when looking at Figure 11, the largest number of projects is in category of sustainable consumption is relevant (81.1%). There are only 18,9% of projects where the sustainable consumption is not relevant.

In the second graph, showing the classification where sustainable consumption is relevant, the biggest part of the projects is in Category C where sustainable consumption is relevant and partly used (54,2%). This means that the projects employ the appropriate technology but they do not think about changing the consumption patterns of citizens. The next category is Category D. This category involved 37.5% of the projects and Category B contains the remaining 8.3%.

Figure 12 shows us the countries locating the projects involved in Category A.

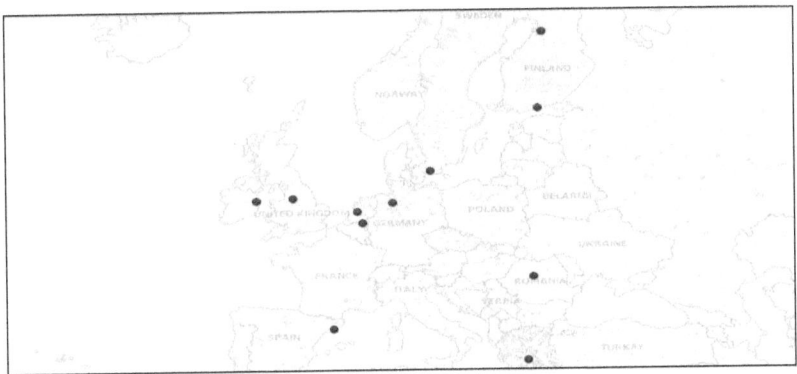

Figure 12: Projects evaluated in Europe with the Category A (SCo not relevant)

These kinds of projects do not hold any relevance about sustainable consumption. These projects are about mobility and real time data, future internet services, digital urban services, mobile applications to live more intelligently, to strengthen the economy for businesses and communities, etc.

Projects that are in Category A are not necessarily bad, but instead are focused on other types of services to citizens. Such projects are good for the progress of improving lives of citizens and for the cities' development. It is very important that these types of projects cover the entire city so that all citizens can take advantage of them.

Figure 13 shows us the projects evaluated involved in the B (SCo is relevant but not present), C (SCo is relevant and partly used) **and** D (SCo is relevant and used to its full potential) **categories of sustainable consumption in Europe.**

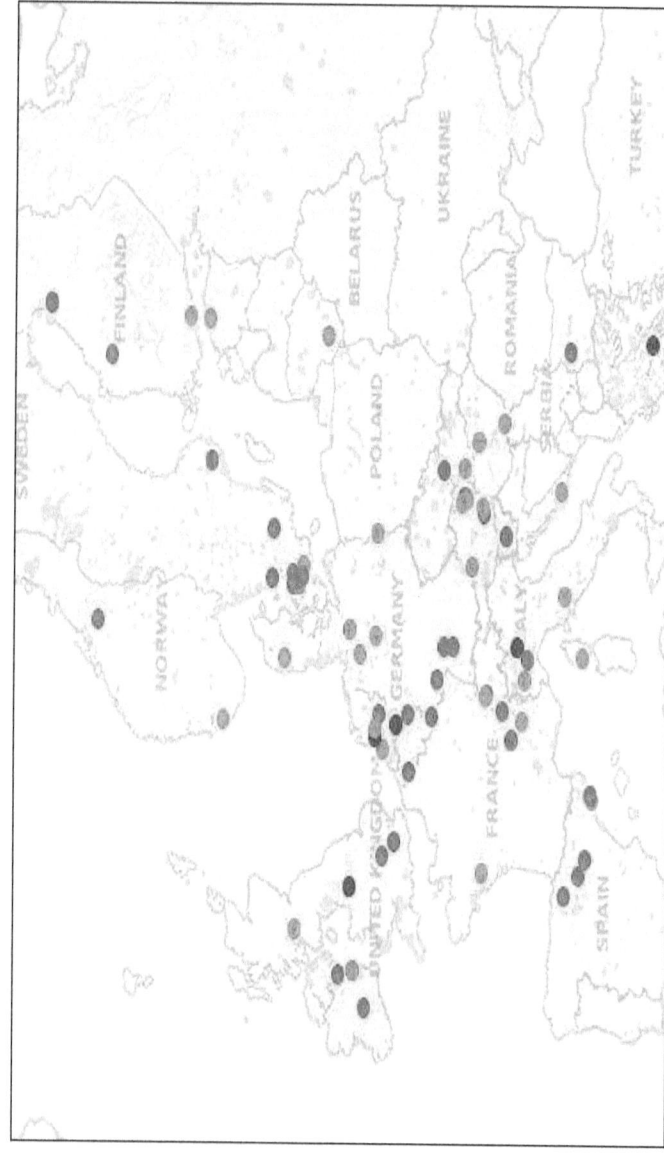

Figure 13: Projects involved in the B, C and D categories of sustainable consumption in Europe

31

According to Figure 13, projects that consider sustainable consumption relevant but not implemented can be found in Category B. It is possible to find projects in this category such as public wifi, data from vehicles, mobile applications, projects to allow citizens to share their experiences of eco-friendly heating services and products, etc.

It is important to question why projects in Category B consider sustainable consumption when the concept is not implemented in the projects. Thus, the companies instigating these projects have a mindset about consumption patterns and they must increase their ambitions for achieving the parameter of sustainable consumption to be reflected in future projects. They could possibly host educational campaigns for more rational use of their products and services and focus on the sustainable use of those.

It can be appreciated that in Category C, most of these projects are involved in energy efficiency and alternative energy sources, ICT solutions to reduce energy consumption, and CO_2 emissions, but the projects do not take into consideration the changes in citizens' lifestyle to consume even less. Projects that are in Category C use the right technology but if they do not care about the behaviour of the citizens, it is possible to have a big rebound effect. It can be quite prejudicial for sustainable consumption patterns.

If the projects involved in Category D are analysed further, it is possible to find projects such as car-sharing, energy saving campaigns, generation of the citizens' own energy, influencing behaviour of citizens through campaigns, monitoring systems in the apartments to let people know how much energy is being used, greater public awareness to reduce consumption at home, incentives for families that consume less and are contributing to sustainable energy use, etc.

The most important points that place projects in Category D are to provide energy consumption information and how to reduce inhabitants' consumption, to have a high percentage of awareness of sustainable consumption and to integrate population in the projects. If the projects follow the guidelines of Category D, the rebound effect will be reduced.

The next table shows the number of projects and also the percentage of the projects in Categories B, C and D per country.

	B (SCo is relevant but not present)		C (SCo is relevant and partly is used)		D (SCo is relevant and used to its full potential)	
	Projects	% projects' country	Projects	% projects' country	Projects	% projects' country
France		0	4	80	1	20
Denmark		0	5	41,6	7	58,4
Ireland	1	20	2	40	2	40
Austria		0	6	75	2	25
Sweden		0	4	40	6	60
The Netherlands	2	18,2	5	45,4	4	36,4
Lithuania		0	1	100		0
Slovakia		0	1	100		0
Germany		0	5	55,5	4	44,4
Croatia		0	1	100		0
Finalnd	1	20	2	40	2	40
United Kingdom	3	30	4	40	3	30
Hungary		0	5	100		0
Switzerland		0	1	33,4	2	66,6
Norway		0	1	50	1	50
Spain			6	54,5	5	45,5
Slovenia		0	2	66,6	1	33,4
Italy	1	11,2	7	77,6	1	11,2
Belgium		0		0	1	100
Luxemburg		0		0	1	100
Poland		0	1	100		0
Bulgaria		0		0	1	100
Czech Republic		0		0	1	100
Estonia		0	2	100		0
Greece	2	100		0		0
Romania		0		0		0

Table 3: Projects' classification by categories of sustainable consumption relevant

According to Figure 13 and Table 3, some interesting aspects across the countries and their projects can be seen:

- **Countries that have some projects in Category B (SCo is relevant but not present):**
  The countries that have projects in Category B are Ireland, Finland, The Netherlands, United Kingdom, Italy and Greece. It is important to look at these countries' projects in depth. Category B is the one where projects are not involved in achieving sustainable consumption patterns. Most of these countries have other projects in Categories C and D but Greece does not have any projects in the higher categories of all the projects evaluated. Projects that are in Category B will have to increase their effort to achieve more ambitious goals. If they stay in this category

they are not going to change their behavioural consumption patterns and reduce the impacts of citizens.

- **Countries with more percentage in Category C (SCo is relevant and part is used):**
The countries that have a significant percentage in Category C are France, Austria, The Netherlands, Germany, United Kingdom, Hungary, Spain, Slovenia, Italy and Estonia. Most of these countries are in central Europe. The projects that these countries have are less ambitious than the countries in the classification below and also most of them only think about technology. They do not consider changing the citizens' behaviour or how they will be involved in the projects to maximize the benefit and to create an awareness of sustainable consumption patterns.

Projects that are in Category C need to be aware that the promotion of energy efficiency is sometimes not the best way to save energy and reduce emissions. The rebound effect can be very high if they do not do a good promotion to reduce energy consumption and CO_2 emissions. It should be taken into consideration that populations do not make the mistake of thinking that through energy savings generated by the new technologies, it is possible to abuse the same product or other products.

- **Countries with greater percentage of projects in Category D (SCo is present and used to its full potential) or the same percentage of projects in Categories C and D:**
The countries in this category are Denmark, Sweden, Switzerland, Finland, Norway and Ireland. Switzerland is the only one that is not in the north of Europe. These countries have a good opportunity to fight Climate Change with ambitious projects, reducing energy consumption and CO_2 emissions. Aside from this, the population of these countries will change their behaviours to follow guidelines of sustainable consumption and to realize the impact of their actions. If future projects follow this mindset, it is possible to achieve the goals of Europe 2020.

4. Discussion and conclusions

According to the definitions of Smart Cities analysed in this work, it is possible to find some common features and one constant aspect in most of the definitions is technology. ICT's are included in projects to improve the adaptability of a city. Nevertheless, the concept of sustainability or environmental responsibility is not sufficiently present in these definitions.

By looking at the definitions of smart sustainable cities, excluding keywords in Smart Cities definitions and how to achieve the objectives through ICTs, other factors come into focus; resource efficiency, cost and energy savings, environmental impacts, etc. According to the definitions of smart sustainable cities described, it can be noticed that the projects evaluated have characteristics mentioned by the definitions. Most of the projects are related using information and communication technologies for a smarter and more efficient use of resources. In addition, they offer citizens better lives, environmental responsibility with a sustainable approach, energy savings and a reduction of the environmental footprint. It is also possible to find some projects (in less quantity), involving citizens in city governance, offering tangible economic growth and employment opportunities for citizens and investing in human and social capital.

The sectors of smart cities in the projects evaluated, mostly include projects that are in the sectors of Smart Environment and Smart Living. These projects integrate technology to reduce consumptions of energy and CO_2 emissions and also integrate energy management systems in buildings and technology to provide living in more comfortable houses. The other sectors that follow Smart Environment and Smart Living are, in decreasing order of their quantity of projects, Smart Citizens, Smart Economy, Smart Mobility and Smart Governance.

In the two databases compared, Concerto's database has many projects in Smart Environment and Smart Living and few projects in the other sectors of smart city. However, the database of Mapping Smart Cities in Europe has a more regular distribution of projects in the different sectors of smart cities. Both databases have more projects in Smart Environment and fewer in Smart Governance.

From the evaluation of the relevance of sustainable consumption in the different sectors of smart cities, it is possible to affirm that sustainable consumption is not relevant in the sectors of Smart Governance and Smart Economy. This is in contrast to the sectors of Smart Citizens, Smart Mobility, Smart Living and Smart Environment, in which the concept is relevant.

From the results of sustainable consumption, 18.9% of the projects evaluated belong to the first classification. Projects involved in this category do not have any relation to sustainable consumption. However, it is important to have these kinds of projects to improve the cities' skills and to give services to citizens.

The other classification is integrated by categories where sustainable consumption is relevant. We saw that 8.3% of the projects are in the lower category of the classification, which means that sustainable consumption is relevant but not implemented in the projects. Projects in this category should be more ambitious to reach a higher level of sustainable consumption. In addition, these projects are intended to change the behavioural consumption patterns, thereby reducing the impact of citizens.

After analyzing the results, it can be seen that most of the projects evaluated consider sustainable consumption guidelines. 54.2% of the projects are in the second category of the SCo relevant. These projects implemented technology to reduce consumption but they do not care about changing the citizens' behaviour. The rebound effect could be high if they do not involve the behaviour of citizens alongside this. The largest amount of these projects is located in central Europe.

Most of the projects that are in the higher category of SCo relevance are in northern Europe. They represent 37.5% of the projects and they make the concept of sustainable consumption to its full potential relevant. Through campaigns and explanations on how to manage and reduce energy consumption, it should be possible to change citizens' behaviour.

It is important to identify some characteristics that allow projects to be in the higher category of SCo relevance. These characteristics are all related with the knowledge and behaviour of citizens. It is remarkable that these types of projects are related to car-sharing, generation of the citizens' own energy, greater public awareness to reduce consumption at home, incentives for families that consume less and are contributing to sustainable energy use. It is noticeable that citizens in this area present sustainable consumption mind sent patterns. Projects involved in energy saving campaigns can also be found, influencing the behaviour of citizens, monitoring systems in apartments to let people know how much energy is spent, etc.

In conclusion, there are few projects that do not take into consideration sustainable consumption relevance. Therefore projects in Europe are moving towards the objectives to reduce CO_2 emissions and energy consumption. Nevertheless, it is relevant to take into consideration the rebound effect that can be seen in several projects, and focus on how to reduce it.

Before beginning to work on this thesis, the main question was if sustainable consumption patterns have been involved in projects and how citizens could contribute to reduce their actions through new technologies. During this project, it has been realized that smart consumption is an important aspect that should be present in smart sustainable cities. Projects have to involve this concept and be sure that citizens are already prepared to cooperate. New technologies help us save energy and reduce our consumption, but if there is a lack of smart consumption from the citizens, the rebound effect will be high and consumption could increase.

It is promising to see that projects are more focused on sustainable consumption and teaching that consumer actions affect the environment. It is undoubtedly necessary to focus on consumer learning because once the agents begin to understand and become aware of sustainable consumption, there will be a tendency to follow and continue on this path.

As the research and analysis of projects were being carried out, we encountered some limitations. Sometimes the information that was extracted from the projects was not clear or lacked clear explanation. In addition, limitations were found when classifying the projects into the different sectors of smart cities and in the classification of sustainable consumption. Sometimes the information of the projects found was not evident, and it would have been better to discuss these classifications with other researchers.

The number of projects evaluated was not the same in every country. For this reason, countries that had only one project were discarded for not being representative. If analysis was conducted with the same number of projects evaluated in each country, the classification of both sectors of smart cities and sustainable consumption would be have been more suitable for comparisons across countries to be made.

Reading the projects of the cities and seinge how high the ambitions of the stakeholders are as well as the classification of the smart sectors and the sustainable consumption after the analysis were the most enjoyable part. Moreover, it was interesting to notice that the results give us an overview of the kind of projects presently occurring in Europe.

Since I grew up in Barcelona I have seen the transformation of the city on how to improve its skills. I have appreciated the development of the city in terms of the sector of smart city. However, nowadays I can understand the objectives of the projects that are involved in Barcelona better and the goals that they want to achieve. There are several examples of projects in my city that I have observed such as Bicing, to improve the green mobility in the city, car sharing, network traffic management, real time data, the district 22@, renewable energies, smart buildings, etc.

After this work, I have noticed that there are other questions to be answered following this thesis. I have made several proposals that can be discussed in future research:

1. Analyze in detail the environmental cost of the technology involved in the project and the benefits obtained.
   Sometimes the cost of new technologies with the goal of reducing energy consumption or emissions is much higher than the benefits obtained. Therefore, the environmental payback has to be taken into consideration. Therefore, a high quantity of natural resources is needed to make good technology. On the other hand, the energy that can be saved with this technology is not high enough.

2. Question the relationship between the ambition of projects and the GDP per capita of the countries.
   It is valuable to know whether countries with a higher GDP per capita have more ambitious projects than the countries with lower GDP per capita.

3. Analyze the success of the projects.
   If the projects have achieved their objectives, how were they achieved? To answer this question, it will be necessary to wait a number of years to be sure that the projects are on a path to achieving their objectives. Additionally, it should be possible to see the limitations of the projects.

5. References

BANBURY, C., STINEROCK, R. and SUBRAHMANYAN, S. (2012) Sustainable consumption: Introspecting across multiple lived cultures. Journal of Business Research, Vol. 65, Issue 4, pp. 497-503.

BRUNTLAND, G.H., 1987. Our Common Future: World Commission on Environment and Development. Oxford University Press, Oxford.

BUENSTORF, G. and CORDES, C. (2008) Can sustainable consumption be learned? A model of cultural evolution. Ecological economics, Vol. 67, Issue 4, pp. 646-657.

BUHAUG, H. and URDAL, H. (2013) An urbanization bomb? Population growth and social disorder in cities. Global Environment Change, Vol. 23, Issue 1, pp. 1-10.

CARLEY, M. and SPAPENS, P. (1998) Sharing the World: sustainable living and global equity in the 21st century. United kingdom: Earthscan Publications Limited, pp. 107-189.

DAMMANN et al. (2005) Energy solutions for smart cities and communities [Online] Available from: http://concerto.eu/concerto/concerto-sites-a-projects/sites-con-sites.html [Accessed 19/04/2014]

DIMITROPOULOS, J. and SORRELL, S. (2008) The rebound effect: microeconomic definitions, extensions and limitations. Ecological Economics, Vol. 65, Issue 3, pp. 636-649.

EGGER, S. (2006) Determining a sustainable city model. Environmental Modelling & Software, Vol. 21, Issue 9, pp. 1235-1246.

GIFFINGER, R., FERTNER, C., KRAMAR, H., PICHLER-MILANOVIC, N. and MEIJERS, E. (2007) European Smart Cities [Online] Available from: http://www.smart-cities.eu/model.html [Accessed 16/04/2014]

HERRING, H. and ROY, R. (2007) Technological innovation, energy efficient design and the rebound effect. Technovation, Vol. 27, Issue 4, pp. 194-203.

HERRING, H. (2006) Energy efficiency – a critical view. Energy, Vol. 31, Issue 1, pp.10-20.

HERRING, H. (2004) Rebound effect of Energy Conservation. Encyclopedia of Energy, pp 237-244.

HIRST, P., HUMMERSTONE, E., WEBB, S., KARLSSON, A., BLIN, A., DUFF, M., JORDANOU, M. and DEAKIN, M. (2012) Joint European support for sustainable investment in city areas [PDF] Available from: http://ec.europa.eu/regional_policy/thefunds/instruments/doc/jessica/jessica_horizontal_study_smart_and_sustainable_cities_en.pdf [Accessed 23/04/2014]

ITU-T Focus Group (WG 1) on Smart Sustainable Cities (2013-2016) Smart Sustainable Cities - An Analysis of Definitions. Telecommunication Standardization Sector. SSC-0100-rev-2.

KRAMERS, A., HÖJER, M., LÖVEHAGEN, N. and WANGEL, J. (2014) Smart sustainable cities – Exploring ICT solutions for reduced energy use in cities. Environmental Modelling & Software. pp. 1-11.

LANGSTON, J. et al. (np) International institute for sustainable development [Online] Available from: http://www.iisd.ca/ [Accessed 21/03/2014]

LINDBERG, F., GRIMMOND, C.S.B., YOGESWARAN, N., KOTTHAUS, S. and ALLENA, L. (2013) Impact of city changes and weather on anthropogenic heat flux in Europe 1995–2015. Urban Climate, Vol. 4, pp. 1-15.

LOREK, S. and FUCHS, D. (2013) Strong sustainable consumption governance – precondition for a degrowth path? Journal of Cleaner Production, Vol. 38, pp. 36-43

MANOOCHEHRI, J. (2002) Post-Rio 'Sustainable Cnsumption´: Establishing coherence and a common platform. Society for International Development, 1011-6370 (200209) 45:3; 47-53; 027166.

MANVILLE, C., COCHRANE, G., CAVE, J., MILLARD, J., PEDERSON, J., THAARUP, R., LIEBE, A., WISSNER, M., MASSIK, R. and KOTTERINK, B. (2014) Mapping smart cities in the EU [PDF] Available from: http://www.europarl.europa.eu/RegData/etudes/etudes/join/2014/507480/IPOL-ITRE_ET(2014)507480_EN.pdf [Accessed 06/05/2014]

MITCHELL, W.J. (2000) E-topia, "Urban life, Jim-but not as we know it". The MIT Press, Cambridge Mass.

NAM, T. and PARDO, T.A. (2011) Conceptualizing smart city with dimensions of technology, people and institutions. 12th Annual International Conference on Digital Government Research Conference: Digital Government Innovation in Challenging Times, pp282-291.

NEIROTTI, P., DE MARCO, A., CORINNA, A. MANGANO, G. and SCORRANO, F. (2014) Current trends in Smart City initiatives: Some stylised facts. Cities, Vol. 38, pp. 25-36.

NYLUND (2010) Energy future of the Stockholm region 2010-2050 The way to reduce climate impact [PDF] Available from: http://www.tmr.sll.se/Global/Dokument/publ/2010/2010_r_energy_future_of_the_stockholm_region_2010-2050.pdf [Accessed 25/04/2014]

PHIPPS, M.,OZANNE, L., LUCHS, M., SUBRAHMANYAN, S., KAPITAN. S., CATLIN, J., GAU, R., NAYLOR, R., ROSE, R., SIMPSON, B., WEAVER, T. (2013) Understanding the inherent complexity of sustainable consumption: A social cognitive framework. Journal of Business Research, Vol. 66, pp.1227-1234.

ROELFSEMA, M., DEN ELZEN, M., HOHNE, N., HOF, A. F., BRAUN, N., FEKETE, H., BOTTCHER, H., BRANDSMA, R. and LARKIN, J. (2014) Are major econnomies on track to achieve their pledges for 2020? An assessment of domestic climate and energy polices. Energy Policy, Vol. 67, pp. 781-796.

ROGERS, R. (1998) Cities for a small planet. Great Britain: Faber and Faber limited, pp. 25-65.

RUESTER, S., SCHWENEN, S., FINGER. M. and GLACHANT, J.M. (2014) Energy Policy, Vol. 66, pp. 209-217.

SANNE, C (2002) Willing consumers – or locked-in? Policies for a sustainable consumption. Ecological Economics, Vol. 42, Issues 1-2, pp. 273-287.

SCHURR, A. (n.d.) IBM Smarter cities [Online] Available from: http://www.ibm.com/smarterplanet/us/en/smarter_cities/overview/ [Accessed 12/4/2014]

SPAARGAREN, G. and MOL, A. (2008) Greening global consumption: Redefining politics and authority. Global Environmental Change, Vol. 18, Issue 3, pp. 350-359.

STEINER, U. and TINGGAARD, G. (2013) Is local participation always optimal for sustainable action? The costs of consensus-building in Local Agenda 21. Journal of Environmental Management, Vol. 129, pp. 266-273.

TOWNSEND, A. (2013) Smart cities: big data, civic hackers and the quest for a new utopia. New York: W.W. Norton and Company, Inc.

UNITED NATIONS (2010) UN World Urbanization Prospects: The 2009 Revision. Highlights. Department of Economic and Social Affairs. New York (2010) [PDF] Available from: http://esa.un.org/unpd/wup/Documents/WUP2009_Highlights_Final.pdf [Accessed 23/03/2014]

UNITED NATIONS SUSTAINABLE DEVELOPMENT (1992) Agenda 21 [Online] Available from:

http://sustainabledevelopment.un.org/content/documents/Agenda21.pdf [Accessed 04/03/14].

WILLIAMS, R. (2010) Global Footprint Network, Advancing the Science of Sustainability [Online] Available from:http://www.footprintnetwork.org/es/index.php/newsletter/bv/august_20_2010 [Accessed 06/03/2014].

Appendix I: Classification of the evaluated projects within smart city sectors and sustainable consumption

n° of project	Country	City	Smart Solutions Technology	S. Governance	S. Mobility	S. Living	S. Environment	S. Society	S. Economy	S. Consumption	Explanation of sustainable consumption grade	Web site
1	France	Ajaccio	Solar ventilation system, photovoltaic panels.			x	x			C	They use the right technology but they do not implement patterns to reduce the consumption of users.	http://www.concerto.eu/concerto/concerto-sites-a-projects/sites-con-sites/sites-search-by-name/sites-crescendo-ajaccio.html
2	France	Lyon	Solar thermal, photovoltaic panels, thermal use of biomass.			x	x			C	They use the right technology but they do not implement patterns to reduce the consumption of users.	http://www.concerto.eu/concerto/concerto-sites-a-projects/sites-con-sites/sites-search-by-name/sites-renaissance-lyon.html
3	France	Nantes	Heat pump, photovoltaic panels and thermal use of biomass.			x	x			C	They use the right technology but the project does not care about the citizens' consumption.	http://www.concerto.eu/concerto/concerto-sites-a-projects/sites-con-sites/sites-search-by-name/sites-act2-nantes.html
4	France	Grenoble	Geothermal energy, biomass, solar thermal panels and hydroelectric power stations.				x			C	They use the right technology but it does not say anything about citizens' sustainable consumption.	http://www.concerto.eu/concerto/concerto-sites-a-projects/sites-con-sites/sites-search-by-name/sites-sesac-grenoble.html
5	France	Lyon	Solar panels, car-sharing, data monitoring, electric vehicles.		X	x	x	x		D	Right technology. The users know how much electricity, water and gas they are using in real time and take action to optimize their consumption.	http://www.economie.grandlyon.com/fileadmin/user_upload/fichiers/site_eco/20121121_gl_lyon_smart_community_dp_en.pdf
6	Denmark	Mabjerg	Biogas production for heat and electricity.				x	x	x	C	Right technology, workshops for farmers on energy management.	http://www.concerto.eu/concerto/concerto-sites-a-projects/sites-con-sites/sites-search-by-name/sites-ecostiler-mabjerg.html
7	Denmark	Helsingborg	Wind energy, solar thermal energy, photovoltaics, micro wind turbines, geothermal energy, biomass heating, biogas heating from waste, polygenerations from biomass and biogas and BTES plant.			x	x			D	Right technology, it has created a network and started defining themes and a work procedure for energy saving campaigns towards end-users in the region.	http://www.concerto.eu/concerto/concerto-sites-a-projects/sites-con-sites/sites-search-by-name/sites-ecocity-helsingborg.html

#	Country	City	Technology				Cat	Description	URL
8	Denmark	Stenlose	Thermal use of biomass, solar thermal, mechanical ventilation with heat recovery, monitoring and targeting.		x		C	Energy saving and renewable energy supply, the concept of sustainable consumption is not exploited in all its potential.	http://www.concerto.eu/concerto/concerto-sites-a-projects/sites-con-sites/sites-search-by-name/sites-class1-stenlose.html
9	Denmark	Hillerod	Solar thermal collectors, photovoltaic, wind energy, heat pumps, low-energy district lighting, intelligent energy management system.		x	x	D	Future supply of low energy houses, much increased awareness in the municipality concerning low energy measures, positive awareness of what it means to be CO2-neutral.	http://www.concerto.eu/concerto/concerto-sites-a-projects/sites-con-sites/sites-search-by-name/sites-sorcer-hillerod.html
10	Denmark	Valby	Photovoltaic, solar thermal, cogeneration from biomass, biogas, geothermal energy, monitoring and targeting.		x	x	D	Low energy solutions and renewable energies, introduction of new and better technologies. The sustainable behavior is present in the project.	http://www.concerto.eu/concerto/concerto-sites-a-projects/sites-con-sites/sites-search-by-name/sites-green-solar-cities-valby.html
11	Denmark	Copenhagen	ICT Mobile access, monitoring and sensors.	x	x	x	D	They are discarding their cars and taking bicycles and public transport as the primary means of transportation.	http://www.cpcleantech.com/media/2113602/integrated%20transport.pdf
12	Denmark	Copenhagen	Intelligent traffic system, monitoring and sensors, e-bikes.		x		D	Strong prioritizing of a bicycle-friendly infrastructure, people are conscious to use bicycles and public transport instead of private transport.	https://subsite.kk.dk/sitecore/content/Subsites/CityOfCopenhagen/SubsiteFrontpage/LivingInCopenhagen/~/media/A6581E08C2EF4275BD3CA1DB951215C3.ashx
13	Denmark	Copenhagen	Monitoring and sensors, data, Utilities - Water.		x	x	C	Use of technology for a port with unpolluted waters, improving the lives of citizens. The port can be used for recreational use.	http://www.dac.dk/en/dac-cities/water/copenhagen-from-sewer-to-harbour-bath/
14	Denmark	Copenhagen	Monitoring and sensors, data, Utilities - Water.		x	x	D	Use of technology to provide high quality tap water and to reduce the plastic bottles consumption.	http://www.ecoinnovation.dk
15	Denmark	Copenhagen	Wind power, smart grid and smart meters, data and monitoring.		x	x	D	To promote investments in wind power, the Government offered tax deductions to the families generating their own energy within their own or the neighboring municipality.	http://www.dac.dk/en/dac-cities/sustainable-cities/all-cases/energy/copenhagen-cities-can-run-on-wind-energy/

#	Country	City	Description				Cat	Summary	URL
16	Denmark	Copenhagen	Geothermal, wind, marine and biomass energy, transportation: metro system, bicycles. New buildings, interconnection with existing systems, utilities infrastructure.	X	X	X	C	It focuses on sustainable energy and new types of energy to create solutions for the new sustainable city district in Copenhagen. The project does not say anything about consumption patterns.	http://www.stateofgreen.com/en/Profiles/E-ON-Nordic/Solutions/Nordhavn-%E2%80%93-A-New-Sustainable-Neighbourhood-in-Cope
17	Denmark	Copenhagen	People network, measurement.		X	X	C	To use ICT as an enabler to significantly reduce energy consumption and CO2 emissions.	http://www.eurocities.eu/eurocities/activities/projects/NiCE-Networking-intelligent-Cities-for-Energy-Efficiency
18	Ireland	Dundalk	Photovoltaic, thermal use of biomass, wind power.		X		D	The project has set ambitious targets of 20% renewable heat, 20% renewable electricity and 40% improvement in energy efficiency of selected buildings. To involve a partnership approach to testing low-cost solutions for influencing behavior and demand reduction.	http://www.concerto.eu/concerto/concerto-sites-a-projects/sites-con-sites/sites-con-sites-search-by-name/sites-holistic-dundalk.html
19	Ireland	North Tipperary	Monitoring and targeting, solar thermal, biomass heating, wind power, photovoltaic.		X	X	D	Monitoring system in each house where electricity, gas and renewables are measured. This helps families know their consumption and change their behavior.	http://www.concerto.eu/concerto/concerto-sites-a-projects/sites-con-sites/sites-con-sites-search-by-name/sites-con-sites-north-tipparary.html
20	Ireland	Dublin	Real time data, mobile devices, road sensors and GPS updates	X			A	The project does not mention the term of sustainable consumption. It is only about the mobility and the real time of the city's buses.	http://www.theguardian.com/local-government-network/2013/jun/05/dublin-city-smart-approach-data
21	Ireland	Dublin	Open data	X	X		A	The project has no relation to sustainable consumption. The project offers public operational data available online for others to research.	http://www.dublinked.net/
22	Ireland	Dublin	Open data, open innovation	X	X	X	C	The project wants to facilitate further innovation in the region, long-term sustainability, enhanced citizen-focussed governance and job creation.	http://digitaldublin.ie/

#	Country	City	Technology				Cat.	Description	URL
23	Ireland	Dublin	4g Wireless network	x		x	B	The project takes into account the sustainability concept but it is not talked about.	http://ec.europa.eu/information_society/apps/projects/factsheet/index.cfm?project_ref=297291
24	Ireland	Dublin	People network, measurement.		x	x	C	To use ICT as an enabler to significantly reduce energy consumption and CO$_2$ emissions.	http://www.eurocities.eu/eurocities/activities/projects/NiCE-Networking-Intelligent-Cities-for-Energy-Efficiency
25	Austria	Mödling	Photovoltaic, solar thermal, thermal use of biomass, energetic use of bio waste, hydropower.		x	x	C	The project wants climatic protection, to foster local energy supply and to implement energy efficiency measures but it does not say anything about the citizens' consumption.	http://www.concerto.eu/concerto/concerto-projects/sites-con-sites/sites-search-by-name/sites-holistic-moedling.html
26	Austria	Hartberg	Photovoltaic, solar thermal, biomass heating, hydropower.		x		C	Significant reduction of energy consumption but it has not taken into account the behaviour of the consumption.	http://concerto.eu/concerto/concerto-sites-a-projects/sites-con-sites/sites-search-by-name/sites-solution-hartberg.html
27	Austria	Tulln	Bio fuel.		x	x	C	The project uses the right technology for energy efficiency but ignores sustainable consumption patterns.	http://concerto.eu/concerto/concerto-sites-a-projects/sites-con-sites/sites-search-by-name/sites-sems-tulln.html
28	Austria	Salzburg	Photovoltaic, solar thermal.		x	x	C	The project uses the appropriate technology, the sustainability concept is integrated, they obtain a reduction of energy consumption but not through the behaviour of citizens.	http://concerto.eu/concerto/concerto-sites-a-projects/sites-con-sites/sites-search-by-name/sites-green-solar-cities-salzburg.html
29	Austria	Weiz	Biomass, solar energy, photovoltaic, hydropower.	X	x	x	D	In this project, the solar thermal system produced more energy than calculated and the use of gas was much lower than expected. But the inhabitants do not change their lifestyle to consume even less.	http://concerto.eu/concerto/concerto-sites-a-projects/sites-con-sites/sites-search-by-name/sites-energy-in-minds-weiz-gleisdorf.html

#	Country	City	Technology				Rating	Description	URL
30	Austria	Vienna	Smart power grids, monitoring and detection.		x		D	The project integrates the concept of sustainability, using the right technology and wants to reduce consumption as much as possible with technology. The system recognizes inefficient consumption patterns and identifies potential opportunities for savings.	http://www.siemens.com/innovation/en/news/2013/e_inno_1319_1.htm
31	Austria	Vienna	People network, measurement.		x	x	C	To use ICT as an enabler to significantly reduce energy consumption and CO2 emissions.	http://www.eurocities.eu/eurocities/activities/projects/NiCE-Networking-intelligent-Cities-for-Energy-Efficiency
32	Austria	Vienna	Urban information models, data smart meters, mobile devices.	x	x	x	C	The aim of the project is the optimization of energy consumption, improving inclusion and mobility of people through the technology.	http://www.iscopeproject.net/
33	Sweden	Växjö	Biomass, geothermal energy, photovoltaics, web-based tools.		x	x	D	Monitoring systems in the apartments to let people know how much energy spent. Greater public awareness to reduce consumption at home. Incentives for families that consume less and are contributing to sustainable energy use.	http://concerto.eu/concerto/concerto-sites-a-projects/sites-con-sites/sites-con-sites-search-by-name/sites-sesac-vaexjoe.html
34	Sweden	Falkenberg	Wind power, solar panels, consumption monitoring.		x		D	A smart box in every flat encouraged tenants to keep track of their energy consumption. About 780 tons per year CO2 saving.	http://concerto.eu/concerto/concerto-sites-a-projects/sites-con-sites/sites-con-sites-search-by-name/sites-energy-in-minds-falkenberg.html
35	Sweden	Malmö	Social Media, apps.		x	x	C	The project takes full account of the views of citizens and supports ecologically intelligent life. The concept of sustainable consumption is not present in all its criteria.	http://www.peripheria.eu/places/malmö

#	Country	City	Topic				Cat	Description	URL
36	Sweden	Malmo	Open data, apps, e-skills.	x			A	The project uses open data to create mobile applications to live more intelligently.	http://www.citadelonthemove.eu
37	Sweden	Malmo	People network, measurement.		x		C	To use ICT as an enabler to significantly reduce energy consumption and CO2 emissions.	http://www.eurocities.eu/eurocities/activities/projects/NiCE-Networking-Intelligent-Cities-for-Energy-Efficiency
38	Sweden	Stockholm	Renewable energy, power grids.		x	x	C	The aim of the project is to reduce the consumption through the right technology.	http://www.tmr.sll.se/Global/Dokument/publ/2010/2010_r_energy_future_of_the_stockholm_region_2010-2050.pdf
39	Sweden	Stockholm	Solar panels, PV cells and small-scale wind power. Monitoring consumption		x	x	D	Visualise energy consumption and provide clearer price signals for more efficient energy consumption.	http://www.tmr.sll.se/Global/Dokument/publ/2010/2010_r_energy_future_of_the_stockholm_region_2010-2050.pdf
40	Sweden	Stockholm	Bio fuels, electric cars.		x	x	D	The project wants to increase the popularity of public transport and to use less energy for travelling in the city. It provides efficient energy for the vehicle fleet.	http://www.tmr.sll.se/Global/Dokument/publ/2010/2010_r_energy_future_of_the_stockholm_region_2010-2050.pdf
41	Sweden	Stockholm	Bio fuels.		x	x	D	Electricity consumption is estimated to be 40 per cent more efficient by 2030 compared with 2006. The project uses the right technology to reduce the consumption.	http://www.tmr.sll.se/Global/Dokument/publ/2010/2010_r_energy_future_of_the_stockholm_region_2010-2050.pdf
42	Sweden	Stockholm	Agriculture and forestry as a supplier of energy. Biogas production.		x	x	D	The project's objective is to reduce the agricultural energy consumption. Fossil-free agriculture. It uses the right technology for less consumption.	http://www.tmr.sll.se/Global/Dokument/publ/2010/2010_r_energy_future_of_the_stockholm_region_2010-2050.pdf

43	Sweden	Stockholm	Sorting and recycling, utilizing the biological waste.	x	x		C	The project uses biological waste to produce biogas. It uses the right technology but it does not care about the consumption patterns.	http://www.tmr.sll.se/Global/Dokument/publ/2010/2010_r_energy_future_of_the_stockholm_region_2010-2050.pdf
44	The Netherlands	Amsterdam New West	Thermal use of biogas, photovoltaic, wind power, waste water heat recovery.	x	x		C	The project uses new technology to reduce CO2 emissions and energy consumption. It ignores promotion of reduced consumption in the citizens' lifestyle.	http://concerto.eu/concerto/concerto-sites-a-projects/sites-con-sites/sites-search-by-name/sites-ecostiler-amsterdam-newwest.html
45	The Netherlands	Delft	Photovoltaic, geothermal energy, hydroelectric power stations.	x	x	x	C	The right technology is used to reduce energy consumption but the citizens are not involved to reduce consumption.	http://concerto.eu/concerto/concerto-sites-a-projects/sites-con-sites/sites-search-by-name/sites-sesac-delft.html
46	The Netherlands	Amsterdam North	Solar thermal.	x	x		D	The project uses the right technology and conducted some interviews to determine the degree of satisfaction and attitudes about energy and environment of the people with the project. This reflects a high percentage of awareness of sustainable consumption.	http://concerto.eu/concerto/concerto-sites-a-projects/sites-con-sites/sites-search-by-name/sites-staccato-amsterdam-noord.html
47	The Netherlands	Almere	Solar panels, PV cells	x	x		C	New technologies to reduce the CO2 emissions but it do not involve sustainable consumption patterns.	http://concerto.eu/concerto/concerto-sites-a-projects/sites-con-sites/sites-search-by-name/sites-crescendo-almere.html
48	The Netherlands	Apeldoorn	PV collectors, combined heat and power plant, biogas generated of the sludge from a waste water treatment plant, monitoring and targeting.	x	x		D	The project's aim is to be carbon neutral by 2020. It provides inhabitants with energy consumption information.	http://concerto.eu/concerto/concerto-sites-a-projects/sites-con-sites/sites-search-by-name/sites-sorcer-apeldoorn.html
49	The Netherlands	Heerlen	Geothermal energy use, photovoltaic.	x	x		D	It uses the right technology. There is education to strengthen responsible and environmentally friendly energy management.	http://concerto.eu/concerto/concerto-sites-a-projects/sites-con-sites/sites-search-by-name/sites-remining-lowex-heerlen.html

#	Country	City	Technology				Grade	Description	URL
50	The Netherlands	Eindhoven	Mobile devices, data from vehicles.	x			B	The project provides traffic patterns to solve problems such as congestion, improvement of traffic flow, but the concept is not relevant.	http://www.nxp.com/news/press-releases/2013/02/dutch-city-region-of-eindhoven-works-with-ibm-and-nxp-to-improve-traffic-flow-and-road-safety.html
51	The Netherlands	Eindhoven	People network, measurement.		x	x	C	To use ICT as an enabler to significantly reduce energy consumption and CO_2 emissions.	http://www.eurocities.eu/eurocities/activities/projects/NiCE-Networking-intelligent-Cities-for-Energy-Efficiency
52	The Netherlands	Amsterdam	Apps, fall detection sensors.	x	x	x	D	The project takes into account the improvement of energy efficiency and it talks about the sustainable consumption of the inhabitants.	http://amsterdamsmartcity.com/projects
53	The Netherlands	Amsterdam	Open data, apps, e-skills.	x		x	A	The project helps to use open data to create mobile applications to live more intelligently. There is nothing related to sustainable consumption.	http://www.citadelonthemove.eu/en-us/home.aspx
54	The Netherlands	Amsterdam	People network, measurement.		x	x	C	To use ICT as an enabler to significantly reduce energy consumption and CO_2 emissions.	http://www.eurocities.eu/eurocities/activities/projects/NiCE-Networking-intelligent-Cities-for-Energy-Efficiency
55	The Netherlands	Amsterdam	Open sensors network, WiFi, open data portal.	x	x	x	A	The project is about applications that make the citizen connected to digital services. Sustainable consumption is not a point in the project.	http://ec.europa.eu/regional_policy/conferences/urban2014/doc/presentations/katalin_gallyas_open_data_commons.pdf
56	The Netherlands	Amsterdam	Public available WiFi	x		x	B	The project highlight is to have a greater connection services. The project thinks about the sustainable consumption but it is not present on it.	http://www.digital-cities.eu
57	The Netherlands	Amsterdam	Open data, open sensor networks.		x	x	A	The project focuses on innovation in public sectors in a scenario of future internet services. The project does not refer to the concept of sustainable consumption.	http://www.opencities.net/node/22

#	Country	City	Technologies				Cat.	Description	URL
58	Lithuania	Bristonas	Solar radiation, biomass, geothermal energy.		x	x	C	The project implements measures to reduce energy consumption through technology but it does not address the consumption of citizens.	http://concerto.eu/concerto/sites-a-projects/sites-con-sites/sites-search-by-name/sites-eco-life-bristonas.html
59	Slovakia	Galanta	Solar radiation, biomass, biogas, geothermal energy.	x	x	x	C	The project improves the energy efficiency through technology but it does not say anything about less consumption or how to get better behaviours.	http://concerto.eu/concerto/sites-a-projects/sites-con-sites/sites-search-by-name/sites-geocom-galanta.html
60	Germany	Hannover	Photovoltaic, geothermal Energy, solar thermal, thermal use of biomass.		x	x	C	The correct technology is used but disregards the behaviour of people. There has not been any kind of campaign to reduce consumption.	http://concerto.eu/concerto/sites-a-projects/sites-con-sites/sites-search-by-name/sites-act2-hannover.html
61	Germany	Ostfildern	Bio energy, geothermal energy, heat pump, photovoltaic, active cooling.		x	x	D	Low CO_2 emissions through the right technology. There are some initiatives related to energy savings in the community.	http://concerto.eu/concerto/sites-a-projects/sites-con-sites/sites-search-by-name/sites-polycity-ostfildern.html
62	Germany	Weilerbach	Wind power, solar thermal, photovoltaic, thermal use of biomass.		x	x	D	The project uses the right technology and promotes the use of technologies to reduce household consumption. Citizens can attend workshops to learn and follow examples.	http://concerto.eu/concerto/sites-a-projects/sites-con-sites/sites-search-by-name/sites-sems-weilerbach.html
63	Germany	Neckarsulm	Solar thermal collectors, photovoltaic, biomass/ wood pellets.		x		D	33.3% of energy savings. Activities to inform the public about energy issues. There was an increase in thought power through awareness campaigns, conferences and energy workshops.	http://concerto.eu/concerto/sites-a-projects/sites-con-sites/sites-search-by-name/sites-energy-in-minds-neckarsulm.html
64	Germany	Hamburg	Electric cars, green building.	x	x		C	The project wants to achieve a smart city through the renewable energies and technology by remodelling existing infrastructures. It uses the right technology but does not think about sustainable consumption.	http://www.dac.dk/en/dac-cities/sustainable-cities-2/all-cases/master-plan/hamburg-hafencity---bringing-the-city-to-yhe-water/?bbredirect=true

#	Country	City	Topic				Cat	Description	URL
65	Germany	Hamburg	Energy efficiency	x	x	x	C	The aim of the project is to achieve energy intelligent cooperation in the various urban districts and to develop innovative energy efficiency services. The project only thinks about the technology used.	http://www.eneff-stadt.info/en/heatingcooling-networks/project/details/intelligent-network-of-urban-infrastructures-smart-power-hamburg
66	Germany	Hamburg	Networks of people, data.		x	x	D	The project offers support to reduce the consumption through targeted exchange and learning activities.	http://www.greendigitalcharter.eu/niceproject
67	Germany	Bremen	Efficiency of heating, real time data, centralised control.	x	x	x	C	The energy consumption is down to 15% to 18% and the buildings in the system are simple and more efficient due to the software. The project does not take into account the consumption patterns.	http://smartcitiescouncil.com/resources/city-bremen-cuts-energy-consumption-and-consolidates-building-management-wonderware-solution
68	Germany	Bremen	Traffic guidance system, real time data, sensors at specific points at the street.	x		x	C	The project involves a traffic management system controlled by sensors, parking management and sensing pollution. It is useful for reducing the pollution and the travel times.	http://peripheria.wikispaces.com/Bremen
69	Germany	Bremen	IP platform, location based services, app.			x	A	The project objective is to improve the quality of life in college through applications and services. The project has no relation to sustainable consumption.	http://www.people-project.eu/portal/index.php?option=com_content&view=article&id=79&Itemid=30
70	Croatia	Hvar	Photovoltaic, solar thermal, hydropower, energetic use of biomass is planned.	x	x		C	The project's objective is to demonstrate the energetic self-sufficiency up to a quota of 20% until 2020. It wants to get advantages from the natural resources to produce energy but it does not think about the behaviour of the citizens and the reduction of their consumption.	http://concerto.eu/concerto/concerto-sites-a-projects/sites-con-sites/sites-con-sites-search-by-name/sites-solution-hvar.html

#	Country	City	Technology					D/A	Comment	URL
71	Finland	Lapua	Biomass heating, solar/ thermal driving cooling, wind power.		x	x	x	D	The project has improved knowledge of energy efficiency, renewable energy possibilities and energy saving through monitoring and automation.	http://concerto.eu/concerto-sites-a-projects/sites-con-sites/sites-con-sites-search-by-name/sites-solution-lapua.html
72	Finland	Helsinki	Mobile devices, real time traffic information.	x	x		x	A	This project is involved in the development of digital urban services that make travelling and living in the city easier. It does not think about how to move in a sustainable way and to reduce the mobility consumption.	http://www.forumvirium.fi/en/project-areas/smart-city
73	Finland	Helsinki	Open democracy, app.	x			x	A	The themes of the project are transparency of the city decision-making and enabling better feedback from the citizens to the civil servants. The concept of sustainable consumption is not present.	http://www.forumvirium.fi/en/project-areas/smart-city/open-helsinki-hack-at-home
74	Finland	Helsinki	Mobile devices, real time data.		x		x	A	The project is involved in the development of digital urban services that make travelling and living in the city easier. Mobile devices are used. It does not reference the sustainable consumption.	http://www.forumvirium.fi/en
75	Finland	Helsinki	Open sensors network, WiFi.	x	x		x	A	Digital services that reach the entire community and encourage innovation and creation. Ignores sustainable consumption.	http://commonsforeurope.net/
76	Finland	Helsinki	Open data, apps, e-skills.	x			x	A	The project helps to use open data to create mobile applications to live more intelligently. It has nothing related to sustainable consumption.	http://www.citadelonthemove.eu/en-us/cities.aspx
77	Finland	Helsinki	Service delivery, e-business.		x	x	x	A	The project involves open data and giving developers the tools they need. The project does not reference the concept of sustainable consumption.	http://www.forumvirium.fi/en/project-areas/smart-city/citysdk

#	Country	City	Topic				Cat	Description	URL
78	Finland	Helsinki	People network, measurement.		x	x	C	To use ICT as an enabler to significantly reduce energy consumption and CO$_2$ emissions.	http://www.eurocities.eu/eurocities/activities/projects/NiCE-Networking-intelligent-Cities-for-Energy-Efficiency
79	Finland	Helsinki	Open sensor networks.	x		x	A	The project focuses on innovation in the public sector in a scenario of future internet services. The project does not refer to the concept of sustainable consumption.	http://www.opencities.net/node/23
80	Finland	Oulu	ICT-Apps, e-skills.	x		x	A	The project aims to enhance business for companies and communities by enabling user-driven planning. It does not refer to the concept of sustainable consumption.	http://www.smart-ip.eu/cases/oulu/
81	Finland	Oulu	Wireless access points, high speed internet.	x		x	A	The project aims to strengthen the economy for businesses and communities but it says nothing about growth in a sustainable way and thinking about the sustainable consumption.	http://www.openlivinglabs.eu/livinglab/oulu-labs-oulu-urban-living-labs
82	Finland	Oulu	Network, communication.	x	x	x	B	The project's aim is to provide technology and design frameworks that can be used for designing and adopting context based services (e.g. instant messaging, local and interoperable services in cities/urban spaces). The concept of sustainable consumption is relevant but it is not present in the project.	http://www.smarturbanspaces.org/partners/i/921/47/city-of-oulu
83	Finland	Oulu	Renewable energy, smart grid.	x	x	x	D	The project's aim is to implement sustainable initiatives for its residents. The project involves citizens and teaches them how to change their habits and follow sustainable consumption patterns.	http://www.cleantechinvestor.com/portal/spotlight/11469-oulu-smart-city.html
84	Finland	Oulu	People network, measurement.		x	x	C	To use ICT as an enabler to significantly reduce energy consumption and CO$_2$ emissions.	http://www.eurocities.eu/eurocities/activities/projects/NiCE-Networking-intelligent-Cities-for-Energy-Efficiency

85	United Kingdom	Milton Keynes	Photovoltaic, heat pumps, monitoring and targeting.		x	x		D	The project uses the technology to change sources of energy. Users can measure their consumption and reduce it.	http://concerto.eu/concerto/concerto-sites-a-projects/sites-con-sites/sites-con-sites-search-by-name/sites-crescendo-milton-keynes.html
86	United Kingdom	Lambeth (London)	Photovoltaic, solar thermal, wind power, thermal use of biogas.		x	x		D	The project uses the right technology and also has an influential position in the population to help them to focus on the path of sustainable consumption. It provides information to the public to reduce their consumption individually.	http://concerto.eu/concerto/concerto-sites-a-projects/sites-con-sites/sites-con-sites-search-by-name/sites-ecostiler-lambeth.html
87	United Kingdom	Manchester	E-service, service delivery.	x	x	x		B	The project thinks about the sustainable consumption but the concept is not present in it.	http://www.epic-cities.eu/content/epic-held-successful-workshop-enoll-event-manchester
88	United Kingdom	Manchester	Mobile technologies, WIFI.	x	x	x		B	The project focuses on the problem of low adoption of ICT by European local authorities in non-metropolitan areas. They take into account sustainable consumption but it is not present in the project.	http://go-on-manchester.com/wordpress/wp-content/uploads/2012/11/Digital-Manchester-info.pdf
89	United Kingdom	Manchester	Internet technologies, open data portal, free Europe WiFi pilot, open sensors network.	x	x		x	A	Digital services that reach the entire community encourage innovation and creation. Ignores sustainable consumption.	http://commonsforeurope.net/
90	United Kingdom	Manchester	Broadband, internet access, e-skills.		x			A	The project involves the regeneration of the most deprived urban areas and it provides access to the internet network in an economical manner. The project does not say anything about sustainable consumption.	http://www.eastserve.com/Nav%20Links/about-us.html

#	Country	City	Keywords				Cat.	Description	URL
91	United Kingdom	Manchester	Eco heating, internet, citizen as a sensor.	x			B	The project allows citizens to share their experiences of eco heating services and products. It takes in account a climate change and carbon reduction. The project thinks about sustainable consumption guidelines but it is not present in it.	http://www.smart-ip.eu/cases/manchester/
92	United Kingdom	Manchester	Internet, e-skills.		x	x	A	The project is a national campaign to promote digital inclusion. The key focus is to bring the benefits of the Internet to all citizens. It says nothing about the sustainable consumption.	http://go-on-manchester.com/
93	United Kingdom	Manchester	E-Business, service delivery.	x		x	A	The project's aim is to improve the delivery of services in urban environments. It is about delivery of services.	http://www.citysdk.eu/manchester-city-council/
94	United Kingdom	Manchester	Energy efficiency.		x		C	Energy efficiency projects using the right technology. It does not take into account citizens' sustainable consumption patterns.	http://www.ireenproject.eu/partners
95	United Kingdom	Manchester	ICT+App, e-skills.			x	A	The project's aim is to stimulate citizen engagement in becoming active generators of content and application development. It does not have any point about sustainable consumption.	http://www.manchesterdda.com/smartip/
96	United Kingdom	Manchester	People network, measurement.		x	x	C	To use ICT as an enabler to significantly reduce energy consumption and CO2 emissions.	http://www.eurocities.eu/eurocities/activities/projects/NiCE-Networking-Intelligent-Cities-for-Energy-Efficiency
97	United Kingdom	Glasgow	Energy efficiency, lighting system.	x	x		C	Glasgow is developing an integrated operations centre; looking at social transport, street lighting, energy efficiency and active travel. The project thinks about the sustainable consumption but not in its totality.	http://sustainablefutures.info/2013/05/02/smart-cities-glasgow-barcelona

#	Country	City	Focus				D	Description	URL
98	United Kingdom	Glasgow	Energy efficiency, transport integration, real time data, mobile access.	x			D	Addressing issues such as energy conservation and generation, greater use of green technology and the integration of active transport routes with public transport networks.	http://futurecity.glasgow.gov.uk/
99	United Kingdom	Glasgow	People network, measurement.		x	x	C	To use ICT as an enabler to significantly reduce energy consumption and CO_2 emissions.	http://www.eurocities.eu/eurocities/activities/projects/NiCE-Networking-Intelligent-Cities-for-Energy-Efficiency
100	Hungary	Mórahalom	Solar radiation, geothermal energy.		x	x	C	The key benefits include lower municipal costs in heating / cooling public buildings, public lighting and better use of geothermal resources. Sustainable consumption patterns from citizens are not present in the project.	http://concerto.eu/concerto-sites-a-projects/sites-con-sites/sites-con-sites-search-by-name/sites-geocom-morahalom.html
101	Hungary	Szentendre	Insulation, monitoring and targeting.		x		C	The solving of energy generation and energy efficiency problems. There is no information about citizen's sustainable consumption.	http://concerto.eu/concerto-sites-a-projects/sites-con-sites/sites-con-sites-search-by-name/sites-pimes-szentendre.html
102	Hungary	Óbuda	Solar thermal, insulation.		x	x	C	Reduction of CO_2 emissions as a result of better energy efficiency and the application of renewable energy sources. Energy savings and, hence, cost savings for inhabitants. The citizens do not follow guidelines for sustainable consumption.	http://concerto.eu/concerto-sites-a-projects/sites-con-sites/sites-con-sites-search-by-name/sites-staccato-obuda.html
103	Hungary	Budapest	Network traffic management.	x		x	C	The aim is to create an urban transport knowledge centre and how to reduce emissions in the city in this sector.	http://www.tide-innovation.eu/en/TIDE-Cities/BKK-Budapest/
104	Hungary	Budapest	People network, measurement		x	x	C	To use ICT as an enabler to significantly reduce energy consumption and CO_2 emissions.	http://www.eurocities.eu/eurocities/activities/projects/NiCE-Networking-Intelligent-Cities-for-Energy-Efficiency

	Country	City	Technologies				Cat.	Description	URL
105	Switzerland	Neuchatel	Photovoltaic, thermal use of biomass, solar thermal, hydro power, monitoring and targeting.	x			D	Energy production and consumption measures that will reduce the use of non-renewable energy by 23%, while substantially increasing renewable energy use. The project is changing the citizen's behaviour.	http://concerto.eu/concerto/concerto-sites-a-projects/sites-con-sites/sites-search-by-name/sites-holistic-neuchatel.html
106	Switzerland	Geneva	Photovoltaic, solar thermal, geothermal energy.	x	x	x	D	The project's aim is a 25% reduction in conventional energy consumption. Quality of information and a general need to communicate to enhance user's behaviour. Communication campaigns and training activities have been determinant in considering Lake Geneva as a natural resource for the city's sustainable development.	http://concerto.eu/concerto/concerto-sites-a-projects/sites-con-sites/sites-search-by-name/sites-tetraener-geneva.html
107	Switzerland	Cernier	Photovoltaic, solar thermal, geothermal energy, biomass heating, wind power, hydropower.	x	x		C	The community's aims are to reach energy independence by using 70% of renewable energy for heating and 90% for electricity. They use the right technology to approach the goals.	http://concerto.eu/concerto/concerto-sites-a-projects/sites-con-sites/sites-search-by-name/sites-solution-cernier.html
108	Norway	Sandnes	Solar energy, insulation, monitoring and targeting.	x	x	x	C	The aim will be a sustainable environment with high technology energy solutions and building materials. It does not use the term of sustainable consumption in its totality.	http://concerto.eu/concerto/concerto-sites-a-projects/sites-con-sites/sites-search-by-name/sites-pimes-sandnes.html
109	Norway	Trondheim	Solar thermal energy, biomass heating.	x	x		D	The project uses the right technology to save energy and there are different campaigns to inform the citizens on how to reduce their consumption.	http://concerto.eu/concerto/concerto-sites-a-projects/sites-con-sites/sites-search-by-name/sites-ecocity-trondheim.html
110	Spain	Tudela	Photovoltaic, wind power, solar thermal energy, geothermal energy, monitoring and targeting.	x	x	x	D	The project involves all the community to participate in it, to demonstrate to people the importance of energy efficiency and that energy efficiency can mean a comprehensive transformation to improve a neighbourhood.	http://concerto.eu/concerto/concerto-sites-a-projects/sites-con-sites/sites-search-by-name/sites-ecocity-tudela.html

#	Country	City	Technologies				Type	Description	URL
111	Spain	Vitoria Gasteiz	Wind energy, solar energy, photovoltaic, geothermal energy, biomass.	x			D	The project is a promotion of subsidized dwellings whose design will take into account criteria of energy efficiency, harnessing of renewable energy and sustainability, encouraging them to participate in the improvement of the quality of life and reduction of the environmental impact.	http://concerto.eu/concerto/concerto-sites-a-projects/sites-con-sites/sites-search-by-name/sites-pimes-vitoria-gasteiz.html
112	Spain	Cerdanyola del Vallès	Photovoltaic, solar heat, wind power.	x			C	The project's aim is to become a model of sustainable growth. They use the right technology to approach the goals.	http://concerto.eu/concerto/concerto-sites-a-projects/sites-con-sites/sites-search-by-name/sites-polycity-cerdanyola-del-valles.html
113	Spain	Zaragoza	Photovoltaic, solar Thermal, thermal use of biomass.	x	x	x	D	The common objective involves a bioclimatic approach of social housing, with high energy-efficiency criteria. It enables advice regarding their energy consumption thanks to the monitoring of a significant share of the constructed or retrofitted dwellings under the project.	http://concerto.eu/concerto/concerto-sites-a-projects/sites-con-sites/sites-search-by-name/sites-renaissance-zaragoza.html
114	Spain	Viladecans	Solar water heater, photovoltaic.	x	x		C	The measures will lead to an energy saving of 30%. The project does not use the term of sustainable consumption in all its totality.	http://concerto.eu/concerto/concerto-sites-a-projects/sites-con-sites/sites-search-by-name/sites-crescendo-viladecans.html
115	Spain	Barcelona	Technical resource, e-skills	x	x	x	C	The project uses the right technology to reduce the energy consumption but it only thinks about the technology.	http://smartcity.santcugat.cat/
116	Spain	Barcelona	Smart grid, LED lighting, micro grids.		x		C	The project aims to upgrade its power supply system in Barcelona where it will roll out a cutting-edge smart grid offering greater savings and more efficient and sustainable management.	http://www.endesa.com/en/aboutEndesa/businessLines/principalesproyectos/Barcelona_Smartcity

#	Country	City	Topic						Cat	Description	URL
117	Spain	Barcelona	Energy measurement, parking space sensors, real time data.	x		x	x		C	The project uses the right technology to reduce energy consumption and implement useful services for the citizens. The concept of sustainable consumption is not totally present in the project.	http://newsroom.cisco.com/press-release-content?articleId=680179
118	Spain	Barcelona	Solar panels	x			x	x	D	People are educated on the use and maintenance of their solar panels. To promote the solar programs and other environmentally sustainable initiatives.	http://www.c40.org/case_studies/barcelonas-solar-hot-water-ordinance
119	Spain	Barcelona	ICT mobile access, promote the bike for transport.		x	x	x		D	To encourage sustainable travel within the city and thus reduce CO2 emissions from transport.	http://www.c40.org/case_studies/bicing-%E2%80%93-changing-transport-modes-in-barcelona
120	Spain	Barcelona	Open data, apps, e-skills	x				x	A	The project helps to use open data to create mobile applications to live more intelligently. It has nothing related to sustainable consumption.	http://www.citadelonthemove.eu/en-us/cities.aspx
121	Spain	Barcelona	People network, measurement.			x	x		C	To use ICT as an enabler to significantly reduce energy consumption and CO2 emissions.	http://www.eurocities.eu/eurocities/activities/projects/NiCE-Networking-Intelligent-Cities-for-Energy-Efficiency
122	Spain	Barcelona	Networks				x	x	A	The project's aim is to improve the cities using information facilities. It offers new services to develop projects. No relation to the concept of sustainable consumption.	http://icityproject.eu/
123	Spain	Barcelona	Service delivery, e-business		x		x	x	A	The project is a national campaign to promote digital inclusion. The key focus is to bring the benefits of the Internet to all citizens. It says nothing about the sustainable consumption.	http://www.citysdk.eu/bcn/

124	Spain	Barcelona	Open sensors network, WiFi	x	x	x	A	Digital services that reach the entire community encourage innovation and creation. Ignores sustainable consumption.	http://commonsforeurope.net/
125	Spain	Barcelona	Open sensor networks		x	x	A	The project focuses on innovation in the public sector in a scenario of future internet services. The project does not refer to the concept of sustainable consumption.	http://opencities.net/barcelona
126	Slovenia	Zagorje	Photovoltaic, solar thermal, geothermal energy, cogeneration plant using biomass.		x		C	The objective is to increase the renewable energy supply by 60% compared to standard national practices. It only involves technology.	http://concerto.eu/concerto/concerto-sites-a-projects/sites-con-sites/sites-con-sites-search-by-name/sites-remining-lowex-zagorje.html
127	Slovenia	Ljubljana	Renewable energy		x	x	D	The project aims to encourage direct communication with citizens on environmental protection, sustainable use of energy and renewable energy. There is an awareness of the population to follow the guidelines of sustainable consumption.	http://www.ljubljanapametnomesto.si/
128	Slovenia	Ljubljana	Electric vehicles, real time information		x	x	C	The project's objectives are innovative strategies in urban traffic that contribute to energy efficiency and alternative energy sources in transport and environmental protection. It does not take into account the consumption patterns.	http://www.civitasljubljana.si/
129	Italy	Alessandria	Photovoltaic, solar thermal, thermal use of biomass.		x	x	D	Design workshops to implement a renewable energy network, monitoring the effectiveness of the steps taken and ensuring that everyone in the district knows about the steps being taken in order to ensure a strong multiplying effect.	http://concerto.eu/concerto/concerto-sites-a-projects/sites-con-sites/sites-con-sites-search-by-name/sites-concerto-al-piano-alessandria.html

#	Country	City	Technology					Grade	Description	URL
130	Italy	Montieri	Solar radiation, geothermal energy, photovoltaic panels, solar thermal collectors.			x		C	The project uses the right technology to reduce the energy consumption but it does not think about the totality of the sustainable consumption concept.	http://concerto.eu/concerto/concerto-sites-a-projects/sites-con-sites/sites-search-by-name/sites-geocom-montieri.html
131	Italy	Turin	Photovoltaic, active cooling (solar/thermal driven cooling), insulation.	x	x	x		C	The project wants to achieve a 46% saving of energy but it does not say anything about sustainable consumption patterns.	http://concerto.eu/concerto/concerto-sites-a-projects/sites-con-sites/sites-search-by-name/sites-polycity-turin.html
132	Italy	Milan	LED street lights, electric car recharging.		x	x		C	The project uses the right technology for the reduction of the emissions (the street lighting brings a saving of 18 tonnes of CO_2 per year) but it does not use the sustainable consumption concept in its totality.	http://www.wantedinmilan.com/news/2002221/milan-s-smart-street-lit-up.html
133	Italy	Milan	Integrated mobility system.	x	x	x		C	The aim of the project is the global optimisation of traffic for better air quality levels. It wants to reduce pollution in the city. They do not think about the sustainable consumption in its totality.	http://www.cost.eu/download/Mori
134	Italy	Milan	Mobile Apps				x	B	The project applies creativity and internet technologies to address sustainability. The concept is relevant but it is not in the project.	http://www.peripheria.eu/places/milan

#	Country	City	Technology			C/D	Objective	URL
135	Italy	Milan	Network traffic management, open access server.	x		C	The objective of the project is the organization of public transport, increasing non-motorized transport and promoting electric vehicles. The concept of sustainable consumption is not fully exploited.	http://www.comune.milano.it/portale/wps/portal/CDM?WCM_GLOBAL_CONTEXT=/wps/wcm/connect/ContentLibrary/ho%20bisogno%20di/ho%20bisogno%20di/progettoeuropeo_tide&categid=com.ibm.workplace.wcm-api.WCM_Category/IT_CAT_Bisogni_55_04/15c365004878dc46b2cfbb789196337 3/PUBLISHED&categ=IT_CAT_Bisogni_55_04&type=content
136	Italy	Milan	Monitoring and sensors, real time data.		x	x	C	The project aims to fully comprehend energy consumption across the city by choosing representative buildings and areas of consumption inside the buildings chosen.	http://www.smartspaces.eu/es/pilot-sites/milan/
137	Italy	Milan	People network, measurement		x	x	C	To use ICT as an enabler to significantly reduce energy consumption and CO2 emissions.	http://www.eurocities.eu/eurocities/activities/projects/NiCE-Networking-intelligent-Cities-for-Energy-Efficiency
138	Belgium	Kortrijk	Photovoltaic, bio fuel.		x	x	D	The project aims to transform a former 'ghetto' into a CO2-neutral neighbourhood. The project uses the right technology to be totally efficient and sustainable.	http://concerto.eu/concerto/concerto-sites-a-projects/sites-con-sites/sites-con-sites-search-by-name/sites-eco-life-kortrijk.html
139	Luxemburg	Redange	Solar power, thermal use of biomass, thermal use of biogas.		x	x	D	The project aims to meet 100% of the energy demand by renewable sources. Citizens follow guidelines for sustainable consumption. The concept is well integrated into society.	http://concerto.eu/concerto/concerto-sites-a-projects/sites-con-sites/sites-con-sites-search-by-name/sites-sems-redange.html
140	Poland	Slubice	Solar Thermal, photovoltaic, thermal use of biomass, wind power.		x	x	C	The general aim of the project is the reduction of energy demand in public buildings by 30% and the reduction of the city energy demand by 10%. It does not care about the sustainable consumption patterns but it uses the right technology to approach the objectives.	http://concerto.eu/concerto/concerto-sites-a-projects/sites-con-sites/sites-con-sites-search-by-name/sites-sems-slubice-pl.html

141	Bulgaria	Sofia	Solar thermal, insulation.		x	x		D	The project shows that citizens have changed their behaviour and they are following guidelines of sustainable consumption to minimize impacts.	http://concerto.eu/concerto/sites-a-projects/sites-con-sites/sites-search-by-name/sites-staccato-sofia.html
142	Czech Republic	Zlin	Solar thermal systems, photovoltaic, biomass, optimized lighting.		x	x		D	To increase energy awareness and motivate people to implement energy saving measures.	http://concerto.eu/concerto/sites-a-projects/sites-con-sites/sites-search-by-name/sites-energy-in-minds-zlin.html
143	Estonia	Tallinn	Smart card system, monitoring.		x		x	C	The city of Tallinn has free public transport. Thus, there is a reduction in private transport and they do not build new roads for more vehicles. The concept of sustainable consumption is integrated but not all its totality.	http://www.eurocities.eu/eurocities/news/Residents-Enjoy-free-public-transport-in-tallinn-WSPO-952EDX
144	Estonia	Tallinn	People network, measurement.			x	x	C	To use ICT as an enabler to significantly reduce energy consumption and CO2 emissions.	http://www.eurocities.eu/eurocities/activities/projects/NiCE-Networking-intelligent-Cities-for-Energy-Efficiency
145	Greece	Athens	Mobile Apps.	x				B	The project applies creativity and internet technologies to address the sustainability. The concept is relevant but it is not in the project.	http://www.peripheria.eu/places/bremen
146	Greece	Athens	4g Wireless network.	x		x		A	The project uses the internet to solve the citizens' problems. The project is not related to the concept of sustainable consumption.	http://ec.europa.eu/information_society/apps/projects/factsheet/index.cfm?project_ref=297291

63

			Public available WiFi.						URL
147	Greece	Athens		x	x	x		B	The project highlights low adoption of ICT in non metropolitan areas and wants a greater connection of services. The project thinks about the sustainable consumption but it is not present on it.	http://www.digital-cities.eu/
148	Romania	Tirgu Mures	E-Services, cloud computing.	x	x	x		A	The project's objectives are the improvement of the quality of life for citizens and the city's economic growth underpinned by innovation and technology. It does not think about the sustainable consumption patterns.	http://epic-cities.eu/content/testimonials

A: Sustainable consumption is not relevant in the project.

B: Sustainable consumption is relevant but not present in the project.

C: Sustainable consumption is relevant and partly used in the project.

D: Sustainable consumption is relevant and used to its full potential in the project.

i want morebooks!

Buy your books fast and straightforward online - at one of the world's fastest growing online book stores! Environmentally sound due to Print-on-Demand technologies.

Buy your books online at
www.get-morebooks.com

Kaufen Sie Ihre Bücher schnell und unkompliziert online – auf einer der am schnellsten wachsenden Buchhandelsplattformen weltweit!
Dank Print-On-Demand umwelt- und ressourcenschonend produziert.

Bücher schneller online kaufen
www.morebooks.de

OmniScriptum Marketing DEU GmbH
Heinrich-Böcking-Str. 6-8
D - 66121 Saarbrücken
Telefax: +49 681 93 81 567-9

info@omniscriptum.de
www.omniscriptum.de

www.ingramcontent.com/pod-product-compliance
Lightning Source LLC
Chambersburg PA
CBHW031539210526
45464CB00003B/1071